Lecture Notes in Earth Sciences 73

W0051433

Springer-Verlag Berlin Heidelberg GmbH

Zdeněk Martinec

Boundary-Value Problems for Gravimetric Determination of a Precise Geoid

With 51 Figures and 3 Tables

 Springer

Author

Zdeněk Martinec
Department of Geophysics
Faculty of Mathematics and Physics
Charles University
V Holešovičkách 2, 180 00 Prague 8, Czech Republic
e-mail: zdenek@hervam.troja.mff.cuni.cz

"For all Lecture Notes in Earth Sciences published till now please see final pages of the book"

Cataloging-in-Publication data applied for

Die Deutsche Bibliothek - CIP-Einheitsaufnahme

Martinec, Zdeněk:
Boundary value problems for gravimetric determination in a precise geoid : with 3 tables / Zdeněk Martinec. - Berlin ; Heidelberg ; New York ; Barcelona ; Budapest ; Hong Kong ; London ; Milan ; Santa Clara ; Singapore ; Paris ; Tokyo : Springer, 1998
    (Lecture notes in earth sciences ; 73)

ISSN 0930-0317

ISBN 978-3-540-64462-0   ISBN 978-3-540-69785-5 (eBook)
DOI 10.1007/978-3-540-69785-5

© Springer-Verlag Berlin Heidelberg 1998
Originally published by Springer-Verlag Berlin Heidelberg New York in 1983

The use of general descriptive names, registered names, trademarks, etc. in this publication does not imply, even in the absence of a specific statement, that such names are exempt from the relevant protective laws and regulations and therefore free for general use.

Typesetting: Camera ready by author
SPIN: 10569179   32/3142-543210 - Printed on acid-free paper

Contents

List of symbols

a_{1m}	Hörmander's constant
A^B	gravitational attraction of spherical Bouguer shell
A^c	gravitational attraction of compensed/condensed masses
A^t	gravitational attraction of topographical masses
δA	direct topographical effect on gravity
b_0	minor semi-axis of the reference ellipsoid best-fitting the geoid
D	depth of compensation
e	first eccentricity
e_0	first eccentricity of reference ellipsoid
E	linear eccentricity of set of confocal ellipsoidal coordinate surfaces $u =$const.
F	free-air reduction
F	hypergeometric function
g	gravity of the Earth
δg	gravity disturbance
Δg^F	free-air gravity anomaly
Δg_ℓ^F	low-frequency part of Δg^F
$\Delta g^{F,\ell}$	high-frequency part of Δg^F
G	Newton's gravitational constant
h	ellipsoidal height
H	topographical height
j_{ref}	cut-off degree of reference potential
j_{max}	cut-off degree of anomalous potential
K	spherical Poisson's kernel
K^{ell}	ellipsoidal Poisson's kernel
K^ℓ	spheroidal Poisson's kernel
ℓ	distance between points (r, Ω) and (R, Ω')
ℓ_0	distance between points (R, Ω) and (R, Ω')
L	distance between points (r, Ω) and (r', Ω')
L_2	space of square-integrable functions on sphere
M	mass of the Earth
M^c	mass of compensation/condensation layer
M^g	masses below the geoid
M^t	topographical masses
N	geoidal height

q_j	truncation coefficients of Poisson's kernel
q_j^ℓ	high-degree part of q_j
P	point on the Earth's surface
P_g	point on the geoid
P_j	Legendre polynomial of degree j
P_{jm}	Legendre functions of the 1st kind
Q	point on reference ellipsoid
Q_{jm}	Legendre functions of the 2nd kind
r	radial distance
r_g	radius of the geoid
R	radius of mean sphere best-fitting the geoid
R_{jk}	Paul's coefficients
S	spherical Stokes's function
S^{elco}	spherical-ellipsoidal Stokes's function
S^{ell}	ellipsoidal Stokes's function
δS	secondary indirect topographical effect on gravity
T	anomalous gravitational potential harmonic outside the Earth
T_ℓ	low-degree reference part of T
T^ℓ	high-degree part of T
T_{jm}	spherical harmonic coefficients of T
T^h	anomalous gravitational potential harmonic outside the geoid
$T^{h,\ell}$	high-degree part of T^h
u	ellipsoidal coordinate
U	normal gravity potential
V	gravitational potential of the Earth
V^B	gravitational potential of spherical Bouguer shell
V^c	gravitational potential of compensated/condensed masses
V^g	gravitational potential of masses below the geoid
V^R	terrain roughness term
V^t	gravitational potential of topographical masses
$V_{jm}^{t,e}$	spherical harmonic coefficients of V^t in the space external to topographical masses
$V_{jm}^{t,i}$	spherical harmonic coefficients of V^t in the space internal to topographical masses
V^ϱ	gravitational potential of compensated masses
V^σ	gravitational potential of condensed masses
V^ω	centrifugal potential
δV	residual topographical potential
δV_ℓ	low-degree part of δV
δV^ℓ	high-degree part of δV
δw	gravity potential disturbance
W	gravity potential of the Earth
W_0	gauge value of gravity potential on the geoid
x, y, z	Cartesian coordinates

Y_{jm}	surface spherical harmonic of degree j and order m
α	azimuth
β	reduced co-latitude
γ	normal/reference gravity
Γ	gamma function
δ_{ij}	Kronecker symbol
ϵ_h	ellipsoidal correction
ϵ_γ	ellipsoidal correction
ζ	Riemann zeta function
ϑ	co-latitude
κ	condition number
λ	longitude
λ_i	eigenvalue
ϱ	density of topographical masses
$\overline{\varrho}$	density ϱ averaged along topographical column
ϱ_c	density of compensated masses
ϱ_0	mean density of topographical masses
$\Delta\varrho_{Moho}$	density contrast at the Moho
σ	density of condensed masses
χ	angular distance in ellipsoidal coordinates
ψ	angular distance in spherical coordinates
ω	angular velocity of the Earth's rotation
Ω	pair of angular spherical coordinates
$\overline{\Omega}$	pair of angular ellipsoidal coordinates
Ω_0	full solid angle
Ω_{ψ_0}	spherical cap of radius ψ_0

Acknowledgements

This work was partly prepared at the Geodetic Institute of the Stuttgart University where I stationed as a Fellow of the Alexander von Humboldt Foundation, and partly at the Geodetic Institute of the Karlsruhe University during my research visit granted by the German Research Foundation. I am grateful to Prof. Erik Grafarend and Prof. Bernhard Heck for their hospitality enjoyed during my stays there. I wish to take this opportunity to thank Prof. Petr Vaníček who acquainted me with problems of the determination of a precise geoid. The continual exchange of opinions with him via electronic mail reminds me still open questions of this topic. My thanks go to Dr. Ctirad Matyska for his help with many aspects of this work. I wish to express my gratitude to Dr. Alfred Kleusberg, Dr. Sun Wenke, Mr. Mehdi Najafi, and Mr. Peng Ong for their round-table debates when I was visiting the Department of Surveying Engineering at the University of New Brunswick. I extend my indebtedness to Mr. Marc Véronneau who provided me with the data of digital terrain model of Canada. The research reported herein was supported by the Grant Agency of the Czech Republic under the grant No. 205/94/0500, the grant No. 205/97/1015, and by NATO linkage grant SA.5-2-05 (CRG. 950754) 1008/94/JARC-501.

Introduction

Whether the geoid can be determined to a 'sufficient' accuracy has been discussed in geodetic circles for many years. The main objection, by those who do not think it can, has always been that the mass density distribution within the Earth will never be known accurately enough to allow us to compute the geoid to a reasonable level of accuracy. This was the main reason that Molodenskij's quasi-geoid and the theory of its determination were preferred for several decades. With the arrival of GPS and its capability to measure ellipsoidal height differences fairly accurate, the discussion on this subject has been renewed. In recent years, different groups have been trying to compute either an accurate geoid or an accurate quasi-geoid (Sjöberg, 1994; Sideris, 1994; Denker et al., 1994). The final goal of these efforts is the determination of a geoid/quasi-geoid with an error of one centimetre or less (Hipkin, 1994).

The geoid over continental areas has recently been determined by two techniques. First, combining GPS positioning with the orthometric heights results in the 'geometrical' geoid whose undulations with respect to the level ellipsoid are given as the differences of the ellipsoidal (GPS-determined) heights minus the orthometric heights (Mainville et al., 1995; Sideris and She, 1995). On the other hand, surface gravity observations supplemented by geodetic levelling can be used for constructing the 'gravimetric' geoid (Vaníček and Kleusberg, 1987; Stewart and Hipkin, 1990; Milbert, 1990; Featherstone, 1992; Forsberg and Sideris, 1993). As a matter of fact, these techniques are by no means independent, since both make a use of a density hypothesis within the Earth. Only the ellipsoidal height differences resulting from GPS positioning on one hand and gravimetric data on the other hand are independently determined. So, we face the fact that the geoid can be determined in two ways, both depending on mass density distribution within the Earth. Hence, there is a hope for a near future that this overdeterminacy will lead to an improvement of density distribution modelling in the uppermost part of the Earth. In order to reach this goal, the geoid should be determined with a 1 dm accuracy (or better) since a change of the density model in the Earth's crust from a commonly accepted constant 2.67 g/cm^3 to a more realistic 3-D density distribution changes the geoid by a few decimetres only (Martinec, 1993).

A magic accuracy of 1 dm in the determination of gravimetric geoid - not yet realizable in a mountainous terrain - requires not only highly accurate surface gravity observations but also accurate theories and corresponding numerical codes for geoid height computations. The last requirement has not been resolved satisfactory yet since existing theories for geoid computations still contain some assumptions which do not allow one to reach the desired accuracy of 1 dm.

The traditional formulation of the problem for geoid height determination over a continental area with non-zero terrain elevations comes out from the fact that heights of the terrain above the geoid and the modulus of the surface gravity are known with a certain accuracy. (This problem can simply be reformulated in the case that the surface gravity potential is given instead of the terrain heights above the geoid.) The effort is to employ Stokes's integral (Stokes, 1849) for determining the gravity potential of the geoid. To be able to use this integral, three requirements must be satisfied: there are no masses outside the geoid so that the anomalous gravitational potential is harmonic outside the geoid, the gravity observations are referred to the geoid, and the geoid is considered as a sphere in a boundary condition for the anomalous gravitational potential. A remove-compute-restore technique is one possibility which, however, satisfies these requirements only approximately:

- Linearize the boundary condition for modulus of gravity on the Earth's surface with respect to a reference gravity potential;
- Linearize boundary condition for gravity potential on the geoid with respect to a level ellipsoid;
- Replace the gravitational effect of topographical masses on gravity at the Earth's surface by that of an auxiliary mass layer/body located on/below the geoid;
- Continue harmonically the linearized boundary functional from the Earth's surface to the geoid;
- Formulate and solve classical Stokes's boundary-value problem on a spherical geoid;
- Restore the gravitational effect of topographical masses.

Unfortunately, a number of theoretical and practical problems appear when this traditional technique for gravimetric geoid determination is used to compute a 'decimetre geoid'. In this thesis, we will attempt to sort out some of these problems and show the ways how to overcome them.

In Chapter 1 we formulate the boundary-value problem for gravimetric determination of the geoid. The boundary condition constraining the solution has not a usual form, because it contains the unknown anomalous potential referred to both the Earth's surface and the geoid coupled by the topographical height. To emphasize the 'two-boundary' character and adopting the terminology introduced by Sansò (1995), this boundary-value problem is called the Stokes two-boundary-value problem. We analyse numerically the solvability of this problem, in particular, we focus on treating the existence of a short-wavelength part of solution. Moreover, we discuss various approximations frequently used in geodesy converting the 'two-boundary' condition to a 'one-boundary' condition only, referred either to the Earth's surface or to the geoid.

In order to ensure the uniqueness of a solution to the Stokes two-boundary-value problem, the first-degree harmonics of an anomalous potential we are looking for must be excluded apriorily from the solution. However, due to the compensation of topographical masses, the first-degree harmonics of the anomalous potential do not, in general, vanish. In Chapter 2 we show the way how to determine these 'forbidden' harmonics.

The effect of topographical masses on geoid height computation is discussed in Chapters 3 to 6. Adopting a spherical approximation of the geoid, we first derive new formulae for the direct and indirect topographical effects. Our expressions are formulated such that the influence of lateral variations of topographical density can be taken into account. As a by-product of our investigations, we proceed to remove the weak singularity of the Newton integral for topographical effects by introducing spherical 'Bouguer' shell. We study the influence of different approximations used in geodesy to approximate the geoid, particularly, we focus on a planar model of the geoid since this may considerably bias results by systematic errors.

The strong gravitational field induced by topographical masses poses another question, namely, which numerical method should be used to compute the gravitational effect of compensated masses. The traditional way is to expand Newton's kernel into a Taylor series, to take only a few first even series terms, and apply the fast Fourier transform to evaluate these terms numerically (e.g., Sideris et al., 1989). A questionable point of this procedure is whether the Taylor series converges or not, and if so, how many terms of the series should be taken into consideration to approximate the gravitational potential of topographical masses with a prescribed accuracy. In Chapter 5 we will find the condition under which the Taylor series converges and will attempt to estimate the size of individual series terms.

So far, all existing formulations of the direct and both indirect topographical effects have assumed that the topographical masses have a homogeneous mass density. The density was considered equal to a mean crustal value of $\varrho_0 = 2.67$ g/cm^3. This appears to be too coarse a model, especially in the vicinity of lakes, such as the Great Lakes in the central part of North America, because of the large difference between the water density and the mean crustal density ϱ_0. It is thus natural to ask whether the density contrast between lake water and surrounding rock is significant enough, and if it is, how this density inhomogeneity influences a precise geoid computation. This question is treated in Chapter 6. We will also consider another type of lateral density inhomogeneity occurring due to a individual geological formation and local geological factors; these may affect the topographical density as much as 10% to 20%. We estimate their impact on geoidal heights for a density pattern of geological structure beneath the Purcell Mountains (the part of the Canadian Rocky Mountains).

Vaníček and Kleusberg (1987) formulated the Stokes boundary-value problem for a higher-degree reference potential. They showed a number of advantages of such a formulation compared to the traditional formulation of the Stokes problem with a reference gravity field generated by a level ellipsoid. For instance, the truncation error of Stokes's integral applied to observed gravity anomaly data reduced to the reference gravity field is significantly smaller in the case when a higher-degree reference potential is employed (also Vaníček and Sjöberg, 1991).

In Chapter 7, we therefore reformulate the Stokes two-boundary-value problem for gravimetric determination of the geoid considering a satellite gravitational model as a reference. We intend to introduce a reference potential such that it does not depend on the way of compensation or condensation of topographical masses but only on the satellite reference model and on the gravitational field induced by topographical masses. The latter contribution to the reference potential is expressed in the form of an ellipsoidal harmonic series and the expansion coefficients are tabled numerically up to degree 20.

In Chapter 8, we investigate the stability of a discrete downward continuation problem for geoid determination when the surface gravity observations are harmonically continued from the Earth's surface to the geoid. The discrete form of Poisson's integral is used to set up the system of linear algebraic equations describing the problem. The posedness of the downward continuation problem is then tested by means of the eigenvalue analysis of this matrix. Numerically, we will treat the discrete downward continuation of gravity in a particularly rugged region of the Canadian Rocky Mountains.

The compensation of topographical masses is a possible way how to stabilize the downward continuation problem as the spectral contents of the gravity anomalies of compensated topographical masses may significantly differ from that of the original free-air gravity anomalies. Using again surface gravity data from the Canadian Rocky Mountains, we will investigate the efficiency of highly idealized compensation models, namely the Airy-Heiskanen model, the Pratt-Hayford model and Helmert's 2nd condensation technique, to dampen high-frequency oscillations of the free-air gravity anomalies.

Stokes's integral for the gravimetric determination of the geoid requires, besides other assumptions, that the gravity anomalies are referred to a sphere. According to Heiskanen and Moritz (1967, sect.2-14), the absolute error introduced by this spherical approximation does not exceed 1 m in terms of geoidal heights. A simple error analysis reveals that this error can even be larger and it may reach several metres. Such an error is unacceptable at a time where a 'decimetre geoid' is the target. In the Chapter 9, we aim at a solution to the Stokes problem for gravity anomalies distributed on an ellipsoid of revolution. The ellipsoidal approximation of the geoid reflects the reality much better than a spherical approximation since the actual shape of the geoid deviates from a level ellipsoid of revolution by 100 m at most. Treating the geoid as a level ellipsoid of revolution in Stokes's two-boundary-value problem may cause the absolute error of at most 2 cm in geoidal heights.

Green's function to the external Dirichlet boundary-value problem for the Laplace equation with data distributed on an ellipsoid of revolution is constructed in Chapter 10. The ellipsoidal Poisson kernel describing the effect of the ellipticity of the boundary on the solution to this boundary-value problem will be expressed in $O(e_0^2)$-approximation as a finite sum of elementary functions which describe analytically the behaviour of ellipsoidal Poisson kernel at the singular point $\psi = 0$. We intend to demonstrate that the degree of singularity of the ellipsoidal Poisson kernel in the vicinity of its singular point is of the same degree as that of the original spherical Poisson kernel.

In Chapter 11, we investigate the boundary-value problem of Stokes's type with ellipsoidal corrections in the boundary condition for anomalous gravity. Green's function approach enables us to avoid applying an iterative approach to solve this type of boundary-value problem since the solution is determined in one step by computing a Stokes-type integral. The question on the convergency of an iterative scheme that has been recommended so far is thus irrelevant.

Spherical harmonic coefficients of the Earth's external gravity field represent important characteristics of the Earth. Satellite tracking data, land-borne gravity observations and altimeter data must be combined to determine these coefficients with a sufficient reliability and resolution. This altimetry-gravimetry problem is governed by the Laplace equation in the space outside the Earth with a mixed type of boundary conditions. It has been proved that the classical least-squares solution to the *continuous* altimetry-gravimetry problem is not stable. On the contrary, Chapter 12 demonstrates that the least-squares solution to the *discrete* altimetry-gravimetry problem does not fail and is stable under satisfying certain conditions.

Most of the theoretical contributions described herein have already been published in the open literature, or the manuscripts describing the contributions have been either accepted or submitted for publication. I believe that this represents the best reviewing process for any research because the reviewing is done by an international group of referees. Thus, I refer to these papers wherever appropriate.

Chapter 1

The Stokes two-boundary-value problem for geoid determination

It is a common belief that, after removing the first-degree spherical harmonics from the gravitational potential, only a regularization of the downward continuation of a high frequency part of the gravity is necessary to guarantee the existence of a unique solution to the boundary-value problem for gravimetric determination of the geoid. In this chapter, we will deal with the original formulation of the problem prior to the downward continuation of gravity. We intend to demonstrate that, besides the spherical harmonics of degree one, the existence of the solution is not also guaranteed for higher-degree harmonics. This lack of an additional guarantee is due to the fact that the input data – the surface gravity and the potential of the geoid – are prescribed on different boundaries.

1.1 Formulation of the boundary-value problem

Let the geocentric radius of the geoid S_g be described by an angularly dependent function $r = r_g(\Omega)$, where (r, Ω) are the geocentric spherical coordinates, i.e., $(r_g(\Omega), \Omega)$ are points lying on the geoid. We will assume that the function $r_g(\Omega)$ is not known. Let $H(\Omega)$ be the height of the Earth's surface above the geoid reckoned along the geocentric radius. Unlike the geocentric radius of the geoid, we will assume that $H(\Omega)$ is a known function. Finally, let the following quantities be given: the gravity $g(\Omega)$ measured on the Earth's surface, the density $\varrho(r, \Omega)$ of the topographical masses (the masses between the geoid and the Earth's surface), and the gauge value W_0 of the gravity potential on the geoid.

Since it is our intention to deal with the gravity field generated by the Earth's internal masses, we make a few simplifying assumptions. First, we assume that the observations of $g(\Omega)$ are corrected for the attraction of the atmosphere and the direct gravitational effect of the other bodies, mainly the moon and the sun. Second, we assume the Earth is a rigid, undeformable body, uniformly rotating (with a constant angular velocity ω) around a fixed axis passing through its centre of mass. This assumption excludes consideration of the indirect gravitational effect of other celestial bodies, such as tidal deformation of the Earth.

Third, as already mentioned, we assume the height $H(\Omega)$ of the Earth's surface S_t above the geoid reckoned along a geocentric radius to be known. It can be defined by an analogous way as the usual orthometric heights, $H := C/\bar{g}_r$, where C is the geopotential number and $\bar{g}_r$ is the mean value of the r-component of gravity along a geocentric radius between the geoid and the Earth's surface. Values of $\bar{g}_r$ can be estimated by a similar procedure as for $\bar{g}$ (Heiskanen and Moritz, 1967, sect.4-4). However, $\bar{g}$ determined by that procedure is a rough estimate of the actual value. Only after finding the geoid with a high accuracy (better than 1 dm), we will be able to improve both $\bar{g}_r$ and $\bar{g}$.

The question we pose is: how to determine the gravity potential $W(r, \Omega)$ **inside** and **outside** the topographical masses and the radius $r_g(\Omega)$ of the geoid? The problem is governed by the Poisson equation with the boundary conditions given on the free boundaries S_t and S_g coupled by means of height $H(\Omega)$:

$$\nabla^2 W = -4\pi G\varrho + 2\omega^2 \qquad \text{outside } S_g , \qquad (1.1)$$
$$|\text{grad } W| = g \qquad \text{on } S_t , \qquad (1.2)$$
$$W = W_0 \qquad \text{on } S_g , \qquad (1.3)$$
$$W = \frac{1}{2}\omega^2 r^2 \sin^2 \vartheta + \frac{GM}{r} + O\left(\frac{1}{r^3}\right) \qquad r \to \infty , \qquad (1.4)$$

where G is the gravitational constant, M is the mass of the Earth, and ϱ is equal to zero outside the Earth. The first-degree harmonics are left out from the potential W because of the geocentric coordinate system.

1.2   Compensation of topographical masses

To be able to work with a harmonic gravitational potential outside the geoid, it is necessary to replace the topographical masses with an auxiliary body situated below or on the geoid and to find the changing the gravitational potential due to this replacement. To start with, let us have a look at the gravitational potential V^t induced by the topographical masses:

$$V^t(r, \Omega) = G \int_{\Omega_0} \int_{r'=r_g(\Omega')}^{r_g(\Omega')+H(\Omega')} \frac{\varrho(r', \Omega')}{L(r, \psi, r')} r'^2 dr' d\Omega' , \qquad (1.5)$$

where Ω_0 is the full solid angle, $d\Omega' = \sin \vartheta' d\vartheta' d\lambda'$, $L(r, \psi, r')$ is the distance between the computation point (r, Ω) and an integration point (r', Ω'),

$$L(r, \psi, r') := \sqrt{r^2 + r'^2 - 2rr' \cos \psi} , \qquad (1.6)$$

and ψ is the angular distance between the geocentric directions Ω and Ω'. The notation $L(r, \psi, r')$ is used to emphasize the fact that L depends only on radial distances r, and r' and the angular distance ψ between the radii of the points. The sum of the centrifugal potential V^ω plus the gravitational potential V^t is an evident quantity which satisfies eqn.(1.1). However, it is a well-known fact the equipotential surfaces of V^t undulate by several hundreds of metres with respect to a level ellipsoid (Martinec, 1991). Thus, it would not be very advantageous to consider only $V^\omega + V^t$ as a particular solution of eqn.(1.1), but another solution of this equation must be added to $V^\omega + V^t$ in order to reduce a large magnitude of potential V^t.

The fact that the known undulations of the geoid are significantly smaller than those induced by potential V^t indicates that there must exist a compensation mechanism which reduces the gravitational effect of topographical masses. This mechanism is probably mainly connected with lateral mass heterogeneities of the crust but also partly with deep dynamical processes (Matyska, 1994). To describe the compensation mathematically, a number of more or less idealized compensation models have been proposed. For the purpose of geoid computation, we may, in principle, employ any compensation model generating a harmonic gravitational field outside the geoid. For instance, the topographic-isostatic compensation models (e.g., Rummel et al., 1988; Moritz, 1990, chapt.8) are based on the compensation by the anomalies of density distribution $\varrho_c(r, \Omega)$ (> 0) in a layer between the geoid and the compensation level $r_g(\Omega) - D(\Omega)$, $D(\Omega) > 0$, i.e., the gravitational potential

$$V^\varrho(r, \Omega) = G \int_{\Omega_0} \int_{r'=r_g(\Omega')-D(\Omega')}^{r_g(\Omega')} \frac{\varrho_c(r', \Omega')}{L(r, \psi, r')} r'^2 dr' d\Omega' , \qquad (1.7)$$

reduces the gravitational effect of topographical masses. Although the way of compensation might originally be motivated by geophysical ideas, it is, from the point of view of our problem, only an auxiliary mathematical operation. We must only express the change in the potential due to this operation to be able to compute the geoid of the original body.

In the limiting case, the topographical masses may be compensated by a mass surface located on the geoid, i.e., by a layer whose thickness is infinitely small. This kind of compensation, called Helmert's 2nd condensation (Helmert, 1884), is described by Newton's surface integral:

$$V^\sigma(r, \Omega) = G \int_{\Omega_0} \frac{\sigma(\Omega')}{L(r, \psi, r_g(\Omega'))} r_g^2(\Omega') d\Omega' , \qquad (1.8)$$

where $\sigma(\Omega)$ is the density of a condensation layer; the density $\sigma(\Omega)$ may be chosen by various ways depending on the manner of approximation used for fitting the topographical potential V^t by the condensation potential V^σ (see Chapters 2 and 3). There is an open question, partly solved in Chapter 8, whether Helmert's 2nd condensation technique, popular recently (e.g., Martinec et al., 1993), is the best way to compensate the gravitational effect of topographical masses, or whether other types of compensation of topographical masses, e.g., isostatic compensations, stabilize the solution to the Stokes boundary-value problem in a more efficient way.

Having introduced a compensation mechanism of the topographical masses, the associated compensation potential V^c approximating the topographical potential V^t reads

$$V^c := V^\varrho \qquad \text{or} \qquad V^c := V^\sigma \qquad (1.9)$$

for the respective isostatic compensation and Helmert's condensation of topographical masses. Finally, a particular solution of Poisson equation (1.1) may be chosen as $V^\omega + \delta V$, where δV is the so-called *residual topographical potential*,

$$\delta V := V^t - V^c , \qquad (1.10)$$

which is the rest of the fit of V^t by V^c.

1.3 Anomalous potential

The gravity potential W can be considered as a sum of the gravitational potential V^g generated by the masses below the geoid, the topographical potential V^t, and the centrifugal potential V^ω:

$$W := V^g + V^t + V^\omega . \qquad (1.11)$$

Inserting from eqn.(1.10) into (1.11) for potential V^t, the gravity potential W becomes

$$W = V^g + V^c + \delta V + V^\omega . \qquad (1.12)$$

Now, let us decompose the gravity potential $V^g + V^c + V^\omega$ into the sum of the (known) normal gravity potential U generated by a level ellipsoid spinning with the same angular velocity as the Earth and an (unknown) anomalous gravitational potential T^h :

$$T^h := V^g + V^c + V^\omega - U . \qquad (1.13)$$

The superscript 'h' emphasizes that T^h approaches the actual anomalous gravitational potential T of the Earth, $T := W - U$. The difference between T and T^h is given by the residual topographical potential δV:

$$\delta V = T - T^h . \qquad (1.14)$$

Throughout the paper we assume that the mass of the level ellipsoid is equal to the mass of the Earth and the mass-center of the level ellipsoid coincides with that of the Earth. Then the potential T does not contain the zero- and first-degree spherical harmonics. Depending on the way of the compensation of the topographical masses, the residual topographical potential δV contains the zero-degree and/or the first-degree spherical harmonics. Due to these facts and eqn.(1.14), the potential T^h includes, in general, both the zero- and first-degree spherical harmonics. The free, non-linear boundary-value problem (1.1)–(1.4) expressed in terms of the anomalous gravitational potential T^h reads:

$$\nabla^2 T^h = 0 \qquad\qquad \text{outside } S_g\,, \qquad (1.15)$$
$$|\text{grad}(U + T^h + \delta V)| = g \qquad\qquad \text{on } S_t\,, \qquad (1.16)$$
$$U + T^h + \delta V = W_0 \qquad\qquad \text{on } S_g\,, \qquad (1.17)$$

together with the asymptotic condition at infinity

$$T^h \sim O\left(\frac{1}{r}\right) \qquad\qquad r \to \infty\,. \qquad (1.18)$$

Taking the first-degree spherical harmonics of potential T^h (and also of the gravity data g) into consideration, the existence and the uniqueness of the solution to the boundary-value problem (1.15)–(1.18) is not guaranteed. We will leave this point open and return to it later after linearizing boundary conditions (1.16) and (1.17).

1.4   Bruns's formula

Let us now treat the boundary condition (1.17). Let P, P_g, and Q be points on the Earth's surface, the geoid, and the level ellipsoid, respectively. Let points P and P_g lie on the same geocentric radius line and point Q be the so called the normal point to P_g. Point Q is established such that (i) the normal gravity potential U at Q is equal to the actual gravity potential W_0 at P_g, and (ii) P_g and Q lie on the same plumb line of the normal gravity field.

Using this notation, the boundary condition (1.17) may be written as

$$U_{P_g} + T^h_{P_g} + \delta V_{P_g} = W_0\,. \qquad (1.19)$$

The normal potential U_{P_g} may be expressed by means of U and its derivatives at the point Q:

$$\begin{aligned}
U_{P_g} &= U_Q + \left.\frac{\partial U}{\partial h}\right|_Q N + \frac{1}{2}\left.\frac{\partial^2 U}{\partial h^2}\right|_Q N^2 + \dots \\
&= U_Q - \gamma_Q N - \frac{1}{2}\left.\frac{\partial \gamma}{\partial h}\right|_Q N^2 + \dots\,,
\end{aligned} \qquad (1.20)$$

where $\partial/\partial h$ is the derivative along the plumb line of the normal gravity field; γ is the normal gravity, $\gamma := |\text{grad}\,U|$, and N is the height of the geoid above the level ellipsoid reckoned along the plumb line of the normal gravity.

Let us estimate the magnitudes of the Taylor series terms in expansion (1.20). Since

$$\frac{\partial \gamma}{\partial h} = -2\frac{\gamma}{r} + O(e_0^2\gamma)\,, \qquad (1.21)$$

where e_0 is the first eccentricity of the level ellipsoid, the terms on the right-hand side of eqn.(1.20) are of orders:

$$U_Q \approx 10^8\,\text{m}^2\text{s}^{-2}\,, \quad \gamma_Q N \approx 10^3\,\text{m}^2\text{s}^{-2}\,, \quad \frac{1}{2}\frac{\partial \gamma}{\partial h}\bigg|_Q N^2 \approx 1.5\times 10^{-2}\,\text{m}^2\text{s}^{-2}\,; \quad (1.22)$$

the geoidal height N has been estimated by 100 metres. Assuming that $U_Q = W_0$ and substituting eqn.(1.20) into (1.19), we get Bruns's formula for the geoidal height N in the form:

$$N = \frac{1}{\gamma_Q}\left(T^h + \delta V\right)\Big|_{P_g} + O(1.5\times 10^{-3}\text{m})\,. \qquad (1.23)$$

Writing approximately

$$\boxed{N \doteq \frac{1}{\gamma_Q}\left(T^h + \delta V\right)\Big|_{P_g}\,,} \qquad (1.24)$$

we can see that an error of Bruns's formula for geoidal height N does not exceed 2 millimetres (see also Martinec, 1990).

The term $T^h_{P_g}/\gamma_Q$ yields the undulations of the so-called *co-geoid* with respect to the level ellipsoid. Note that the co-geoid is the equipotential surface of the gravity potential $V^g + V^c + V^\omega$ with the gauge value W_0. The term $\delta V_{P_g}/\gamma_Q$ yields the undulations of the geoid with respect to the co-geoid; it is termed the *primary indirect topographical effect on the geoid* (Heck, 1993; Martinec and Vaníček, 1994a).

1.5   Linearization of the boundary condition

Let us apply the operator grad$\{\cdot\}$ to the gravity potential $W = U + T^h + \delta V$ and take the magnitude of the resulting vector. We get

$$|\text{grad}W| = |\text{grad}U|\left[1 + 2\frac{\text{grad}U\cdot\text{grad}(T^h + \delta V)}{|\text{grad}U|^2} + \frac{|\text{grad}(T^h + \delta V)|^2}{|\text{grad}U|^2}\right]^{1/2}\,,$$
$$(1.25)$$

where '$\cdot$' denotes the scalar product of vectors. Employing the binomial theorem, the last formula may be linearized as

$$|\text{grad}W| = |\text{grad}U|\left[1 + \frac{\text{grad}U\cdot\text{grad}(T^h + \delta V)}{|\text{grad}U|^2} + O\left(\frac{|\text{grad}(T^h + \delta V)|}{|\text{grad}U|}\right)^2\right]\,.$$
$$(1.26)$$

The magnitudes of the quadratic and higher-order terms can be estimated as follows

$$\left(\frac{|\text{grad}(T^h + \delta V)|}{|\text{grad}U|}\right)^2 \approx \left(\frac{300\,\text{mgal}}{10^6\,\text{mgal}}\right)^2 \approx 10^{-7}\,. \qquad (1.27)$$

Hence, neglecting the non-linear term in eqn.(1.26) makes a relative error of 10^{-7} and an absolute error of 0.1 mgal, which is the accuracy limit of our theory. Considering such an approximation, eqn.(1.26) reads

$$g = \gamma + \frac{\gamma \cdot \operatorname{grad}(T^h + \delta V)}{\gamma} , \tag{1.28}$$

where $g := |\operatorname{grad} W|$ is the actual gravity. Taking spherical approximation of the second term on the right-hand side corrected to the flattening of the Earth, we approximately get

$$g \doteq \gamma - \frac{\partial T^h}{\partial r} - \frac{\partial \delta V}{\partial r} + \epsilon_h(T^h) + \epsilon_h(\delta V) , \tag{1.29}$$

where the ellipsoidal correction term $\epsilon_h(T)$ is given by (e.g., Jekeli, 1981)

$$\epsilon_h(T) := e_0^2 \sin \vartheta \cos \vartheta \, \frac{1}{r} \frac{\partial T}{\partial \vartheta} . \tag{1.30}$$

Note that eqn.(1.29) is valid everywhere above the geoid.

Particularly, let us consider eqn.(1.29) at a point P on the Earth's surface, we have

$$\left. \frac{\partial T^h}{\partial r} \right|_P - \epsilon_h(T_P^h) = -g_P + \gamma_P - \delta A + \epsilon_h(\delta V_P) , \tag{1.31}$$

where

$$\delta A := \left. \frac{\partial \delta V}{\partial r} \right|_P \tag{1.32}$$

is the *direct topographical effect on gravity* (Heck, 1993; Martinec and Vaníček, 1994b).

Normal gravity γ_P may be expressed by means of γ and its derivatives at the point Q:

$$\gamma_P = \gamma_Q + \left. \frac{\partial \gamma}{\partial h} \right|_Q N + \left. \frac{\partial \gamma}{\partial r} \right|_Q H + \frac{1}{2} \left. \frac{\partial^2 \gamma}{\partial r^2} \right|_Q H^2 + \dots \tag{1.33}$$

where the magnitude of terms neglected does not exceed 0.01 mGal. Introducing the free-air reduction

$$F := - \left. \frac{\partial \gamma}{\partial r} \right|_Q H - \frac{1}{2} \left. \frac{\partial^2 \gamma}{\partial r^2} \right|_Q H^2 - \dots , \tag{1.34}$$

eqn.(1.33) reads

$$\gamma_P = \gamma_Q - F + \left. \frac{\partial \gamma}{\partial h} \right|_Q N . \tag{1.35}$$

The derivative of the normal gravity γ along the plumb line of the normal gravity field can be approximately expressed as (e.g., Cruz, 1986)

$$\frac{\partial \gamma}{\partial h} \doteq -\frac{2\gamma}{r} + \frac{\gamma}{r} e_0^2 \left(3 \cos^2 \vartheta - 2 \right) . \tag{1.36}$$

Substituting eqns.(1.35) and (1.36) into eqn.(1.31) and expressing the geoidal height N by Bruns's formula (1.24), we get the final form of the linearized boundary condition for the potential T^h:

$$\boxed{\left.\frac{\partial T^h}{\partial r}\right|_P + \frac{2}{r_Q}T^h_{P_g} - \epsilon_h(T^h_P) - \epsilon_\gamma(T^h_{P_g}) = -\Delta g^h \,,} \tag{1.37}$$

where the gravity anomaly Δg^h consists of terms

$$\Delta g^h := \Delta g^F + \delta A + \delta S - \epsilon_h(\delta V_P) - \epsilon_\gamma(\delta V_{P_g}) \,. \tag{1.38}$$

Here we have introduced the free-air gravity anomaly Δg^F,

$$\Delta g^F := g_P - \gamma_Q + F \,, \tag{1.39}$$

the ellipsoidal correction term $\epsilon_\gamma(T)$,

$$\epsilon_\gamma(T) := e_0^2 \left(3\cos^2\vartheta - 2\right) \frac{T}{r_Q} \,, \tag{1.40}$$

and the *secondary indirect topographical effect on gravity* δS (Wichiencharoen, 1982),

$$\delta S := \frac{2}{r_Q}\delta V_{P_g} \,. \tag{1.41}$$

It should be pointed out that the error of the linearization of boundary condition (1.37) has not exceeded 0.1 mGal. Thus, the magnitude of the terms which are neglected in the linearized boundary condition (1.37) is 0.1 mGal at most. Seitz et al. (1994) showed that this linearization error produces an error of at most 2 cm in geoidal heights.

Inspecting boundary condition (1.37), we can see that the term $\partial T^h/\partial r$ is referred to the Earth's surface whereas the potential T^h in the term $2T^h/r$ is referred to the geoid. Hence, eqn.(1.37) represents a non-standard boundary condition with the unknown referred to the two boundaries, the geoid and the Earth's surface, coupled by height $H(\Omega)$. Adopting the terminology introduced by Sansò (1995), such a problem can be classified as the pseudo-boundary value problem; here we will call it the Stokes two-boundary-value problem. This terminology emphasizes that instead of having the boundary condition relating either to the Earth's surface or to the geoid, the boundary condition (1.37) contains a mixture of unknown T^h referred to the geoid as well as to the Earth's surface (Martinec and Matyska, 1997).

1.6 The first-degree spherical harmonics

In the contrast to the Stokes problem, the non-standard character of the boundary condition (1.37) may cause the first-degree harmonics of the potential T^h could be determined from the condition (1.37) once $H(\Omega) > 0$. However, the transfer function between these harmonics and the gravity anomalies on the right-hand side of eqn.(1.37) takes values which are much smaller than the values of transfer function for other harmonics of T^h. Thus, the inversion of the transfer function, which includes the first-degree harmonics, is numerically unstable or even does not exist once $H(\Omega) = 0$.

To define the unique and stable solution, we exclude, by definition, the first-degree harmonics of T^h from the solution. The asymptotic condition (1.18) imposed on the potential T^h is then replaced by the constraint of the same form as for the Stokes problem (e.g., Sansò, 1981), i.e.,

$$T^h = \frac{c}{r} + O\left(\frac{1}{r^3}\right) \qquad \text{for } r \to \infty , \qquad (1.42)$$

where c is a constant. Imposing such a uniqueness constraint on the solution implies that the first-degree harmonics of the right-hand side of the boundary condition (1.37) must vanish in order to get a solution, i.e.,

$$\int_{\Omega_0} \Delta g^h(\Omega) Y_{1m}^*(\Omega) d\Omega = 0 \qquad \text{for } m = -1, 0, 1 , \qquad (1.43)$$

where $Y_{1m}(\Omega)$ are spherical harmonics of the first-degree and order m, and asterisk denotes complex conjugate. Such conditions are not, in general, satisfied for experimentally determined function Δg^h. In order to balance the number of unknowns and conditions, we will follow Hörmander's trick (Hörmander, 1976) and modify the boundary condition (1.37) by introducing three extra unknowns a_{1m}, $m = -1, 0, 1$:

$$\left.\frac{\partial T^h}{\partial r}\right|_P + \frac{2}{r_Q} T^h_{P_g} - \epsilon_h(T^h_P) - \epsilon_\gamma(T^h_{P_g}) = -\Delta g^h + \sum_{m=-1}^{1} a_{1m} Y_{1m}(\Omega) . \qquad (1.44)$$

The unknowns constants a_{1m} can be determined from the existence condition (1.43), now applied to the modified boundary condition (1.44). We readily obtain

$$a_{1m} = \int_{\Omega} \Delta g^h(\Omega) Y_{1m}^*(\Omega) d\Omega \qquad \text{for } m = -1, 0, 1 . \qquad (1.45)$$

Summarizing, in order to get a unique and stable solution, the Stokes two-boundary-value problem should be formulated as follows: find the potential T^h outside and on the geoid S_g and three constants a_{1m} such that

$$\nabla^2 T^h = 0 \qquad\qquad \text{outside } S_g, \quad (1.46)$$

$$\left.\frac{\partial T^h}{\partial r}\right|_P + \frac{2}{r_Q} T^h_{P_g} - \epsilon_h(T^h_P) - \epsilon_\gamma(T^h_{P_g}) = -\Delta g^h + \sum_{m=-1}^{1} a_{1m} Y_{1m}(\Omega) , \quad (1.47)$$

$$T^h = \frac{c}{r} + O\left(\frac{1}{r^3}\right) \qquad \text{for } r \to \infty \quad (1.48)$$

together with eqn.(1.24). As we will see in the next section, this formulation still not guarantee the well-posedness of the Stokes two-boundary-value problem.

1.7 Numerical investigations

The original problem (1.1)–(1.4) as well as the problem described by eqns.(1.46)–(1.48) and (1.24) are scalar non-linear free boundary-value problems since the radial coordinate of the geoid is one of the unknowns to be determined. Having some approximation of geoid, it is easy to transform the latter free boundary-value problem to a problem with fixed boundaries. For example, replacing P_g by r_Q, r_Q being the radius of the normal point Q, and P by $r_Q + H(\Omega)$ in eqn.(1.47) yields the ellipsoidal approximation of the Stokes two-boundary-value problem, where eqns.(1.46)–(1.48) serve to determine T^h; eqn.(1.24) then gives the geoidal height N. Another possibility, most often used in geoid height computations, is to approximate the geoid in the boundary condition (1.47) by a mean sphere with radius $R = 6371$ km. This means the radius of the point P_g is replaced by R and radius of the point P by $R + H(\Omega)$. The relative error introduced by this spherical approximation is of the order of 3×10^{-3} in the classical problems (Heiskanen and Moritz, 1967, sect.2-14), which then causes a long-wavelength error of at most 0.5 metres in geoidal heights. In regional problems, where only shorter wavelengths are to be determined, this approximation is often reasonable. In the following numerical tests we will employ the spherical approximation of boundary condition (31) for its simplicity. We intend to concentrate on the effects connected with the 'two-boundary nature' of this condition that appear only in a very short wavelength part of the solution.

The solution to the Laplace equation (1.46) with the condition (1.48) can be represented as a series of solid spherical harmonics $r^{-j-1}Y_{jm}(\Omega)$,

$$T^h(r,\Omega) = \sum_{\substack{j=j_{ref}+1 \\ j \neq 1}}^{j_{max}} \sum_{m=-j}^{j} T_{jm} \left(\frac{R}{r}\right)^{j+1} Y_{jm}(\Omega) , \qquad (1.49)$$

where $j_{ref}(\geq 0)$ and j_{max} are the respective reference and maximum cut-off degrees, $Y_{jm}(\Omega)$ are spherical harmonics of degree j and order m, and T_{jm} are the coefficients of potential T^h to be determined. In order to normalize the potential coefficients T_{jm}, we have introduced the mean Earth's radius R into the expansion (1.49). Equation (1.47) in the spherical approximation then becomes

$$\frac{1}{R} \sum_{\substack{j=j_{ref}+1 \\ j \neq 1}}^{j_{max}} \sum_{m=-j}^{j} \left[(j+1)\left(\frac{R}{R+H(\Omega)}\right)^{j+2} - 2 + e_0^2(3\cos^2\vartheta - 2) \right] Y_{jm}(\Omega)T_{jm} +$$

$$+ \frac{e_0^2}{R} \sum_{\substack{j=j_{ref}+1 \\ j \neq 1}}^{j_{max}} \sum_{m=-j}^{j} \left(\frac{R}{R+H(\Omega)}\right)^{j+2} \sin\vartheta\cos\vartheta \frac{\partial Y_{jm}(\Omega)}{\partial\vartheta} T_{jm} = \qquad (1.50)$$

$$= \Delta g^h - \sum_{m=-1}^{1} a_{1m}Y_{1m}(\Omega) ,$$

where coefficients a_{1m} are given by eqn.(1.45). This boundary condition must hold in any direction Ω. In order to ensure it, we will employ the Galerkin method (Lapidus and Pinder, 1982) in which eqn.(1.50) can be rewritten as a system of linear algebraic equations for coefficients T_{jm}:

$$Am = d \,, \tag{1.51}$$

where m is a column vector composed of potential coefficients T_{jm}, i.e.,

$$m := \{T_{jm}|j = j_{ref} + 1, ..., j_{max}, j \neq 1, m = -j, ..., j\} \,, \tag{1.52}$$

A is the matrix composed of the weighted left-hand side of eqn.(1.50),

$$A_{j_1 m_1, jm} :=$$

$$:= \int_{\Omega_0} \left[(j+1) \left(\frac{R}{R+H(\Omega)} \right)^{j+2} - 2 + e_0^2 (3\cos^2 \vartheta - 2) \right] Y_{jm}(\Omega) Y_{j_1 m_1}^*(\Omega) d\Omega +$$

$$\tag{1.53}$$

$$+ e_0^2 \int_{\Omega_0} \left(\frac{R}{R+H(\Omega)} \right)^{j+2} \sin \vartheta \cos \vartheta \frac{\partial Y_{jm}(\Omega)}{\partial \vartheta} Y_{j_1 m_1}^*(\Omega) d\Omega \,,$$

and d is a column vector of weighted right-hand side of eqn.(1.50),

$$d_{j_1 m_1} := R \int_{\Omega_0} \Delta g^h(\Omega) Y_{j_1 m_1}^*(\Omega) d\Omega \,, \tag{1.54}$$

where $j_1 = j_{ref} + 1, ..., j_{max}, j_1 \neq 1$, and $m_1 = -j_1, ..., j_1$.

1.7.1   An example: constant height

Let us first consider a simple, but illustrative, case when $H = H_0 = const.$ over the Earth, and $e_0^2 = 0$. Introducing function

$$K_j(H_0) := (j+1) \left(\frac{R}{R+H_0} \right)^{j+2} - 2, \quad \text{for } j \geq 2 \,, \tag{1.55}$$

the transfer matrix $A_{j_1 m_1, jm}$ between unknown parameters T_{jm} and the gravity anomalies Δg^h on the right-hand side of eqn.(1.50) becomes $A_{j_1 m_1, jm} = K_j(H_0) \delta_{jj_1} \delta_{mm_1}$ and thus

$$T_{jm} = \frac{R}{K_j(H_0)} \int_{\Omega_0} \Delta g^h(\Omega) Y_{jm}^*(\Omega) d\Omega \,. \tag{1.56}$$

Since $0.998 < R/(R + H_0) < 1$ for the Earth, it is clear that $\lim_{j \to \infty} K_j = -2$ for any fixed $H_0 > 0$. On the other hand, $K_j > 0$ for low degrees j because $0.976 < K_2 < 1$. This means that there is a range of j's in which K_j is zero or near zero. For those j's the solution of eqns.(1.51) is unstable or even does not exist once $K_j = 0$.

Let us estimate the range of j's for which the solution of eqns.(1.51) becomes unstable for this simple example. Figure 1.1 plots the values of K_j for height H_0 equal to 1 km, 5 km and 10 km. We can see that the increase of K_j with increasing j is confined to low degrees j and then K_j starts to decrease to its limiting value -2. That is why, the determination of potential T^h is stable only in some part of the spectral domain. The width of the stable part grows with decreasing H_0.

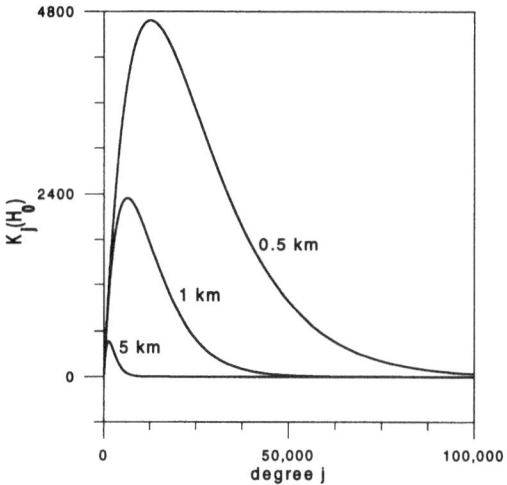

Figure 1.1: Transfer function $K_j(H_0)$ between unknown coefficients T_{jm} and gravity anomalies Δg^h for $H_0 = 1$ km, 5 km, and 10 km.

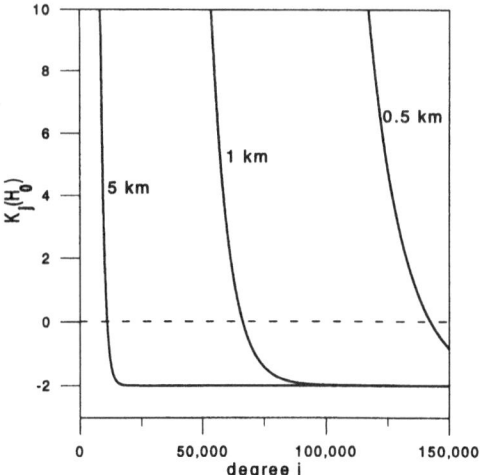

Figure 1.1a: A detail of Figure 1.1.

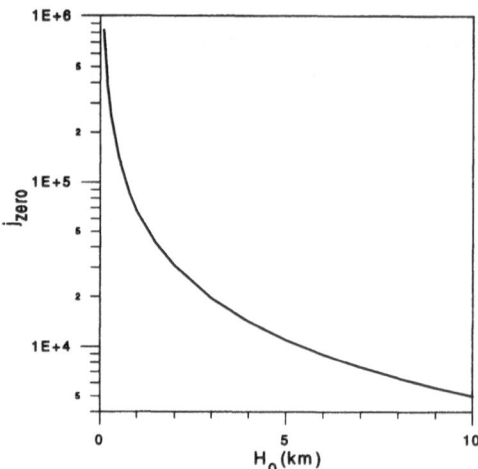

Figure 1.2: The roots j_{zero} of function $K_j(H_0)$ for $H_0 \in (100\,\text{m}, 10^4\,\text{m})$.

Figure 1.2 plots those j_{zero} for which function $K_j(H_0)$ vanishes. For such degrees matrix $\boldsymbol{A}$ is singular and the solution of system of equations (1.51) does not exist. Since spherical degree j corresponds to a given resolution $\Delta\Omega$ in a spatial domain, $\Delta\Omega = \pi/j$, we may also convert critical degree j_{zero} to a critical spatial resolution size $\Delta\Omega_{zero}$, $\Delta\Omega_{zero} = \pi/j_{zero}$, for which the solution to our problem does not exist. Figure 1.2 shows that, for instance, $j_{zero} = 10980$, for $H_0 = 5$ km, and the critical spatial resolution size is $\Delta\Omega_{zero} \doteq 1$ arcmin. To interpret the result in other words, let us imagine that the Earth's topography is a Bouguer spherical shell with a constant height of 5 km above the geoid and the Stokes two-boundary-value problem is solved in a spatial domain such that the potential $T^h(R, \Omega)$ is parameterized by discrete values $T^h(R, \Omega_i)$ in a regular angular grid with grid step size $\Delta\Omega$. Then the solution to the Stokes two-boundary-value problem will not exist if the grid step size $\Delta\Omega$ of the parameterization of T^h is less than or equal to the critical step size $\Delta\Omega_{zero}$, i.e., of about 1 arcmin in our example, even though the surface gravity data would be known continuously on the Earth's surface.

To map the non-existence of the solution for regional geoid determination and for a more realistic model of the Earth's topography, we need to set up and to solve the system of eqn.(1.51) for high degrees and orders ($j_{max} = 10^4 - 10^5$). This leads to computational difficulties because of huge consummation of computational time and memory; with today's computer equipment it is impossible to carry out the analysis of the existence for such a general case. Thus, we are forced to approximate the Earth's surface by a simplified model of axisymmetric geometry. By making use of the analysis of this simplified case, we will attempt to estimate the range of critical spectral degrees j_{zero} for the actual case.

1.7.2 Axisymmetric geometry

Let the height $H(\vartheta, \lambda)$ of the Earth's surface above the geoid is modelled by zonal as well as tesseral and sectoral spherical harmonics of the global digital terrain model TUG87 (Wieser, 1987) cut at degree 180. To create a rotational symmetric body, axisymmetric height $H(\vartheta)$ will be generated by height $H(\vartheta, \lambda)$ taken along a fixed meridian $\lambda = \lambda_0$. In the case of an axisymmetric surface, the elements $A_{j_1 m_1, jm}$ of matrix $\boldsymbol{A}$ do not depend on angular orders m and m_1; they can be written as

$$A_{j_1 j} = \int_{\vartheta=0}^{\pi} \left[(j+1) \left(\frac{R}{R + H(\vartheta)} \right)^{j+2} - 2 + \right.$$

$$\left. + e_0^2 (3 \cos^2 \vartheta - 2) \right] P_j(\cos \vartheta) P_{j_1}(\cos \vartheta) \sin \vartheta \, d\vartheta + \qquad (1.57)$$

$$+ e_0^2 \int_{\vartheta=0}^{\pi} \left(\frac{R}{R + H(\vartheta)} \right)^{j+2} \sin \vartheta \cos \vartheta \, \frac{dP_j(\cos \vartheta)}{d\vartheta} P_{j_1}(\cos \vartheta) \sin \vartheta \, d\vartheta \ .$$

Note that the elements $A_{j_1 j}$ can only be evaluated by a method of numerical quadrature.

To analyse the posedness of the Stokes two-boundary-value problem, we will employ the eigenvalue analysis of matrix $\boldsymbol{A}$. According to this method, a non-symmetric matrix $\boldsymbol{A}$ can be decomposed to the product of three matrices,

$$\boldsymbol{A} = \boldsymbol{U \Lambda U}^{-1} \ , \qquad (1.58)$$

where the columns of matrix $\boldsymbol{U}$ are formed from the right eigenvectors of $\boldsymbol{A}$, the rows of $\boldsymbol{U}^{-1}$ are formed from the left eigenvectors of $\boldsymbol{A}$, and the diagonal matrix $\boldsymbol{\Lambda}$ consists of eigenvalues of $\boldsymbol{A}$. We have employed subroutines BALANC, ELMHES and HQR (Press et al., 1992, sect.11.5 and 11.6) to find the eigenvalues of a non-symmetric matrix $\boldsymbol{A}$.

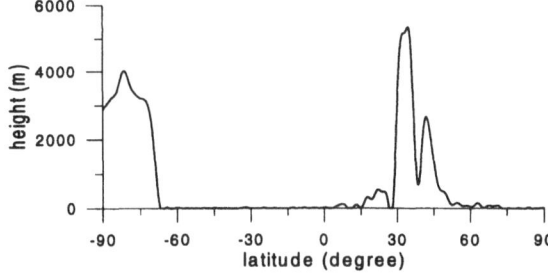

Figure 1.3: The meridian profile $\lambda = 80°$ of topographical height $H(\vartheta, \lambda)$ generated by the global digital terrain model TUG87 (Wieser, 1987) cut at degree 180. This profile is used to create a body with the axisymmetric geometry of external surface.

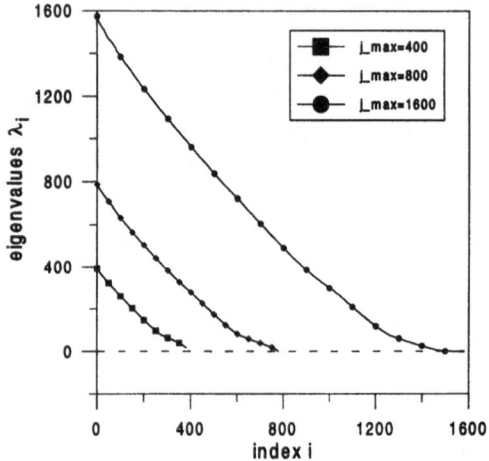

Figure 1.4: The eigenvalue spectra of matrix A for various cut-off degrees j_{max} and a body with axisymmetric surface generated by height $H(\vartheta, \lambda = 80°)$ multiplied by 10 ($j_{ref} = 20$). The ellipsoidal corrections ϵ_h and ϵ_γ are equal to zero.

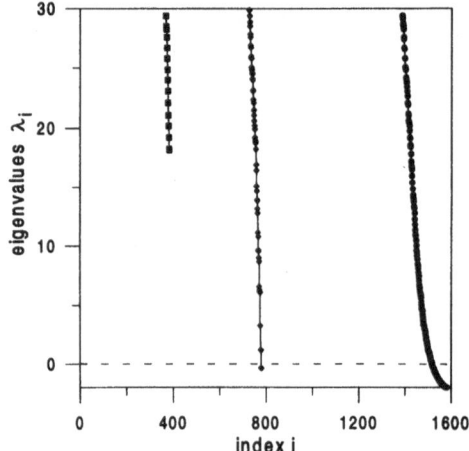

Figure 1.4a: A detail of Figure 1.4.

Figure 1.3 shows the topographical height $H(\vartheta, \lambda_0)$ along the meridian profiles $\lambda_0 = 80°$ reaching value $H_{max} = 5353$ metres. The consequent Figure 1.4 shows a plot of the eigenvalues of matrix A for an axisymmetric body with the outer surface generated by this meridian profile. In order to avoid high degrees j, and thus, be able to perform the eigenvalue analysis in real CPU time, we multiply function $H(\vartheta)$ by a factor of 10. The reference spherical degrees j_{ref} of the potential series (1.49) is $j_{ref} = 20$, which models the situation when low-degree harmonics of potential T^h are determined by another approach, e.g., when considering a satellite gravitational model. In Figure 1.4, where we further put the eccentricity of the level ellipsoid equal to zero, $e_0 = 0$, we change the maximum cut-off degree j_{max} of the potential T^h and plot eigenvalues of matrix $\boldsymbol{A}$ ordered according to their size (note that the eigenvalues are real numbers in this particular case). Inspecting Figure 1.4 we can observe that the eigenvalue spectrum of matrix $\boldsymbol{A}$ intersects the zero level starting from degree $j_{zero} \doteq 800$. Once the cut-off degree j_{max} of the spherical harmonic expansion (1.49) of potential T^h is greater or equal to j_{zero}, the eigenvalue spectrum of $\boldsymbol{A}$ contains a null eigenvalue or an eigenvalues of a very small size. The matrix $\boldsymbol{A}$ becomes ill-conditioned or even singular and the inverse $\boldsymbol{A}^{-1}$ may be distorted by large round-off errors or may not exist at all; in such a case the Stokes two-boundary-value problem does not have a unique and stable solution. As in the preceeding section, the critical degree j_{zero} can again be converted to the critical spatial discretization size $\Delta\Omega_{zero}$ for a case when the Stokes two-boundary-value problem is solved in a spatial domain.

The next test investigates the influence of the ellipsoidal corrections terms ϵ_h and ϵ_γ on the posedness of matrix $\boldsymbol{A}$. We choose the same body as in the preceeding example together with $j_{ref} = 20$ and $j_{max} = 1600$ and compute the eigenvalues of matrix $\boldsymbol{A}$ putting $e_0^2 = 0$ and $e_0^2 = 0.006694$, respectively. Figure 1.5 shows those eigenvalues the magnitudes of which are smaller than 3. (Note that the eigenvalues of $\boldsymbol{A}$ for the case $e_0^2 = 0.006694$ are complex numbers.) We can observe that the eigenvalue spectrum of $\boldsymbol{A}$ changes significantly when e_0^2 differs from zero: there is no null eigenvalue and the magnitude of the smallest eigenvalue is larger than 1. In other words, the ellipsoidal corrections ϵ_γ and ϵ_h

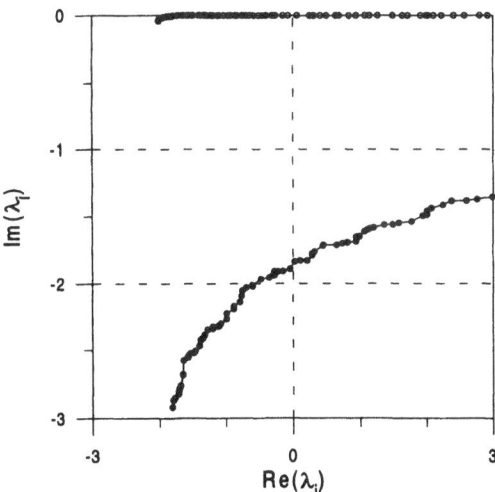

Figure 1.5: The real vs. imaginary parts of the eigenvalues of matrix $\boldsymbol{A}$ with (lower branch) and without (upper branch) the ellipsoidal corrections ϵ_h and ϵ_γ. The axisymmetric body is the same as that considered in Figure 1.4 ($j_{ref} = 20$, and $j_{max} = 1600$).

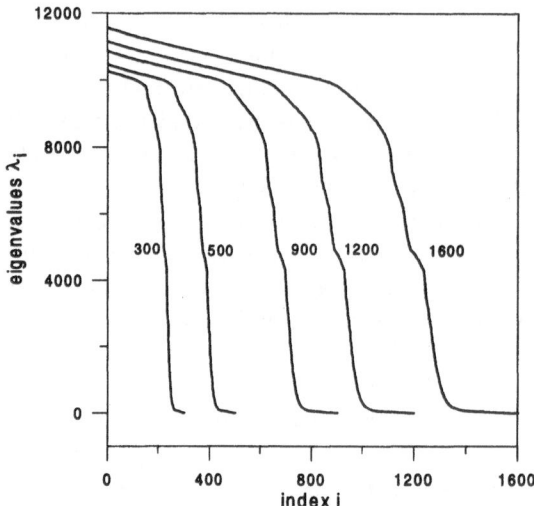

Figure 1.6: The eigenvalue spectra of matrix A for various cut-off degrees $j_{max} = j_{ref} + \Delta j$, $\Delta j = 300, 500, ..., 1600$, and a body with axisymmetric surface generated by height $H(\vartheta, \lambda = 80°)$ ($e_0^2 = 0$ and $j_{ref} = 10000$).

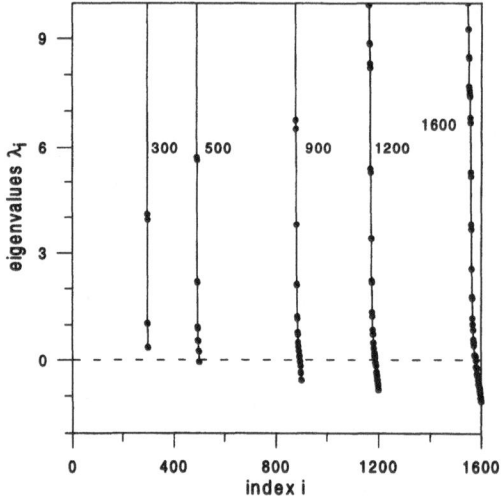

Figure 1.6a: A detail of Figure 1.6.

act as regularization factors removing the ill-posedness of matrix A. It also means that, in this particular case, ϵ_γ and ϵ_h cannot be subtracted from the right-hand side of eqn.(1.51) as known quantities determined a priorily by using a known global gravitational model of the Earth; such usage of ellipsoidal corrections is recommended in real geoid computations (e.g., Cruz, 1985).

. In order to create a more realistic example, we use the same profile of topographical height as plotted in Figure 1.3, but now, in contrast with preceeding example, we will not multiplied height $H(\vartheta)$ by 10. In this case, it is not possible to carry out the eigenvalue analysis of matrix A starting from degree $j_{ref} = 20$ and going up to degrees $j_{max} \approx 10^4 - 10^5$ due to a huge consummation of computer time and memory. We have to confine ourselves to a smaller range of sought spherical harmonics. That is why we choose $j_{ref} = 10000$ and j_{max} in the range between 10300 and 11600. The results for the case $e_0^2 = 0$ are shown in Figure 1.6. We can again observe that eigenvalue spectra intersect the zero-level starting at degree $j_{zero} \doteq 10500$. It means that whenever $j_{max} \geq j_{zero}$, the spectrum of matrix A contains an eigenvalue which is very close or equal to zero. Consequently, matrix A becomes ill-conditioned or even singular. Putting $e_0^2 = 0.006694$ (this case is not plotted here) has a similar stabilization effect as in the case shown in Figure 1.5.

To carry out the eigenvalue analysis of matrix A needs a lot of computer time. However, the critical spherical degree j_{zero} for which the existence of the solution to the Stokes two-boundary-value problem is not guaranteed can be estimated by analysing the existence of a solution for a model with a constant topographical height over the world. If we replace H_0 in the example in section 1.7.1 with the maximum topographical height H_{max}, then such an estimate j_{const} obviously underestimates the actual j_{zero}, i.e., it is too pessimistic, and hence it holds

$$j_{zero} \geq j_{const} , \tag{1.59}$$

where j_{const} is determined by the roots of function $K_j(H_{max})$ given by eqn.(1.55), i.e., j_{const} satisfies the equation

$$(j_{const} + 1) \left(\frac{R}{R + H_{max}} \right)^{j_{const}+2} - 2 = 0 . \tag{1.60}$$

For the examples in Figures 1.4 and 1.6, we obtain $j_{const} \doteq 698$ when $H_{max} = 53530$ metres, and $j_{const} \doteq 10158$ when $H_{max} = 5353$ metres. We have already learnt that the actual critical numbers are $j_{zero} \doteq 800$ and $j_{zero} \doteq 10500$, respectively. So, the criterion (1.59) estimates j_{zero} quite well.

1.8 Different approximations leading to the fundamental equation of physical geodesy

There are at least three approximate approaches how to convert the Stokes two-boundary-value problem to the regular Stokes problem. All of them try to express the 'two-boundary' condition (1.47) in terms of a condition referred to one boundary only, the Earth's surface or the geoid. A common and also an easy way, used, e.g., by Vaníček and Kleusberg (1987), is based on the belief that the approximation

$$\left. \frac{\partial T^h}{\partial r} \right|_P \doteq \left. \frac{\partial T^h}{\partial r} \right|_{P_g} \tag{1.61}$$

does not generate large errors in resulting geoidal heights. The solution to the Stokes problem is then readily to be found by employing Stokes's integration. However, Vaníček et al. (1996) showed that the approximation (1.61) may cause systematic errors in geoidal heights in magnitudes of several decimetres.

The second possibility consists in developing the radial derivative of potential T^h into a Taylor series:

$$\left.\frac{\partial T^h}{\partial r}\right|_P = \left.\frac{\partial T^h}{\partial r}\right|_{P_g} + \left.\frac{\partial^2 T^h}{\partial r^2}\right|_{P_g} H + \ldots . \tag{1.62}$$

Taking only the first two terms of this Taylor series expansion, and putting approximately $\epsilon_h(T_P^h) \doteq \epsilon_h(T_{P_g}^h)$, boundary condition (1.47) takes the form

$$\left.\frac{\partial T^h}{\partial r}\right|_{P_g} + \frac{2}{r_Q}T_{P_g}^h = -\Delta g^h - g_1 + \epsilon_h(T_{P_g}^h) + \epsilon_\gamma(T_{P_g}^h) + \sum_{m=-1}^{1} a_{1m} Y_{1m}(\Omega) , \tag{1.63}$$

where

$$g_1 := \left.\frac{\partial^2 T^h}{\partial r^2}\right|_{P_g} H . \tag{1.64}$$

Now, the two requirements of Stokes's integration are satisfied: the boundary condition (1.63) is referred to a point on the geoid, and the potential T^h is harmonic outside the geoid. Therefore, Stokes's integration may be immediately applied to eqn.(1.63).

However, the right-hand side of eqn.(1.63) contains the term g_1 which makes the problem difficult. Since $\partial^2 T^h/\partial r^2$ is the vertical gradient of the anomalous gravitation $\partial T^h/\partial r$, the term $g_1 = (\partial^2 T_{P_g}^h/\partial r^2) H$ represents the harmonic downward continuation of the anomalous gravitation from P to P_g. Since the actual gravity field is not known in topographical masses, the term g_1 can only be evaluated approximately or the boundary-value problem formulated above must be solve iteratively starting with some model for g_1. The latter approach leads, in fact, to the downward continuation of gravity from the Earth's surface to the geoid. Such a procedure is unstable and requires some kind of regularization to suppress amplifications of short wavelengths (see Chapter 8), however, there are then no problems with the existence of a solution. Hence, approximation (1.62) changes the nature of the problem—it requires a more or less sophisticated procedure in preparing the right-hand side of the boundary condition (1.63) from the observed data instead of dealing with the solvability of the boundary-value problem formulated for original observations; the possibility of the non-existence of the solution is lost by this approximation.

It should be noted that Molodenskij et al. (1960) and later Moritz (1980a, sect.45) made the suggestion to evaluate the term g_1 synthetically in order to avoid the problem with the non-stability of the downward continuation procedure. They assume that the gravity anomalies are linearly dependent on topographical heights and the Poisson integration for the term g_1 is taken over topographical heights and not over gravity data. However, the linear relationship between free-air gravity anomalies and topographical heights introduced by Pellinen (1962) holds only approximately (Heiskanen and Moritz, 1967, Figure 7-6). The question as to the errors of geoidal heights due to this approximation still remains open. The main advantage of the approximation (1.63) is that the solution to the problem (1.46), (1.48) and (1.63) in the spherical approximation may easily be expressed by means of Stokes's integration.

Perhaps the least drastic approximation is to refer the second term on the left-hand side of eqn.(1.47) to the Earth's surface. Formally, boundary condition (1.47) may be rewritten in the form

$$\left.\frac{\partial T^h}{\partial r}\right|_P + \frac{2}{r_P}T^h\bigg|_P = -\Delta g^h - DT^h + \epsilon_h(T_P^h) + \epsilon_\gamma(T_{P_g}^h) + \sum_{m=-1}^{1} a_{1m}Y_{1m}(\Omega) \,, \quad (1.65)$$

where

$$DT^h := \frac{2}{r_g}T^h\bigg|_{P_g} - \frac{2}{r_P}T^h\bigg|_P \,. \quad (1.66)$$

Let us make an estimate of the maximum of DT^h,

$$\left|DT^h\right| \doteq \frac{2}{r_P}\left|\frac{\partial T^h}{\partial r}\right|_P H \leq 200\frac{H}{r_P}\,\text{mGal} \leq 0.25\,\text{mGal} \,, \quad (1.67)$$

where the gravity disturbances $\partial T^h/\partial r$ on the Earth's surface has been estimated by the value of 100 mGal, and the height H of the Earth's surface above the geoid by 8900 metres. In most practical applications, term DT^h may be neglected because its maximum size is less then the accuracy of gravity data available for geoid determination. When the accuracy of gravity anomalies is better than 0.25 mGal, then the term DT^h may be computed from existing models of the geoid, or the Stokes problem with approximate boundary condition (1.65) may be solved iteratively starting with $DT^h=0$, and improving it successively.

One possible way to solve this problem consists of two steps (Martinec, 1996; Vaníček et al., 1996). First, harmonic function $r\partial T^h/\partial r + 2T^h$ is continued from the Earth's surface to the geoid, and then Stokes's integral is employed to find the potential T^h on the geoid. The downward continuation of a harmonic function is an unstable procedure meaning that the solution exponentially diverges at infinite frequency. (To get a bounded, non-oscillating solution, some kind of regularization must be applied to the problem.) Once the boundary condition (1.47) is approximated by the boundary condition (1.65) and the problem is solved as outlined, the property of the problem on the existence is again changed. Whereas the solution to the original Stokes two-boundary-value problem does not exist for a finite geoidal wavelength, the solution to the approximate problem with boundary condition (1.65) exists except for the geoidal wavelength of an infinitesimally short wavelength that must be completely suppressed prior to downward continuation.

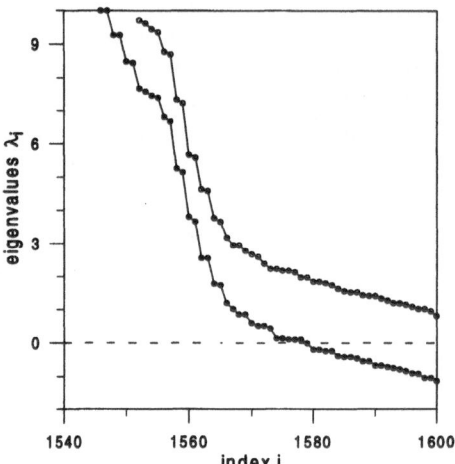

Figure 1.7: The eigenvalue spectra of matrix A for boundary condition (1.47)
(lower branch), and approximate boundary condition (1.65) (upper branch); (the
body is the same as that considered in Figure 1.6; $j_{ref} = 10000$, and $j_{max} = 11600$).

It is possible to avoid the downward continuation and to deal directly with
(1.65). The main disadvantage of this approach is that the boundary operator
is referred to the Earth's surface which cannot be approximated by a smooth
boundary. That is why a simple analytical formula solving (1.46), (1.48), and
(1.65) is not available. Nevertheless, the nature of the problem (1.46), (1.48),
(1.65) is again different from the nature of the original Stokes two-boundary-
value problem. Figure 1.7 demonstrates this fact in a transparent way. For the
axisymmetric example shown in Figure 1.6, we compare the eigenvalue spectrum
of matrix A for the original Stokes two-boundary-value problem to that of the
Stokes problem with approximate condition (1.65). Whereas the first eigenvalue
spectrum intersects the zero level at a finite harmonic degree $j_{zero} \doteq 10500$, the
second spectrum approaches zero for $j \to \infty$.

1.9   Conclusion

This chapter formulated and discussed the existence of a solution to the Stokes
two-boundary-problem for geoid determination. We derived the boundary condition (1.47) relating to this problem without assuming that the surface gravity
data had been continued from the Earth's surface to the geoid. The boundary
condition (1.47) has not a usual form, because it contains the unknown anomalous
gravitational potential referred to both the Earth's surface and the geoid coupled
by the known topographical height. The numerical analysis of the 'two-boundary'
condition (1.47) performed for a simplified model of the Earth's surface has revealed that the transfer matrix between the unknown potential on the geoid and

the surface gravity anomalies may become ill-conditioned or even singular at a certain critical wavelength of a **finite** length. The existence of solution is not guaranteed for this critical geoidal wavelength. Once this ill-posed case occurs, to obtain a bounded and non-oscillating solution, the Stokes two-boundary-value problem must be regularized in such a way that this critical geoidal wavelength and its vicinity are excluded from the solution. We have given an estimate of critical geoidal wavelength; for the highest part of the Earth's surface, the critical geoidal wavelength is about 1 arcmin.

Furthermore, we discussed three possibilities of transforming the original form of 'two-boundary' condition into a 'one-boundary' condition relating to the geoid or the Earth's surface. The approach, based on the Taylor series expansion of the radial derivative of unknown potential, leads to the downward continuation of a harmonic function from the Earth's surface to the geoid the solution of which is unstable for an infinitesimally short geoidal wavelength. It means that the domain of solvability of Stokes two-boundary-value problem is changed by the Taylor series expansion; the solution becomes unstable for shorter geoidal wavelengths compared to that of the original Stokes two-boundary-value problem.

Chapter 2

The zeroth- and first-degree spherical harmonics in the Helmert 2nd condensation technique

In section 1.6 we have excluded, by definition, the first-degree spherical harmonics from the anomalous gravitational potential T^h in order to ensure the uniqueness of the solution to the Stokes two-boundary-value problem. However, when topographical masses are compensated by an auxiliary body consisting of masses below and/or the geoid, the potential T^h may include the zeroth- and first-degree spherical harmonics. In this chapter we will show the way how to determine these 'forbidden' harmonics for Helmert's 2nd condensation technique and estimate their magnitude. Note that similar considerations can be performed for other type of the compensation of topographical masses.

2.1 Principle of mass conservation

Let us first consider the case when surface density $\sigma(\Omega)$ of the Helmert condensation layer is chosen according to principle of mass conservation. This principle postulates that the mass M^c of the condensation layer is equal to the topographical masses M^t, i.e.,

$$M^c = M^t .\tag{2.1}$$

The integral form of this condition reads:

$$\int_{\Omega_0} \sigma(\Omega) r_g^2(\Omega) d\Omega = \int_{\Omega_0} \int_{r=r_g(\Omega)}^{r_g(\Omega)+H(\Omega)} \varrho(r,\Omega) r^2 dr d\Omega .\tag{2.2}$$

The condition (2.2) is satisfied identically if we, for instance, put

$$\sigma(\Omega) = \frac{1}{r_g^2(\Omega)} \int_{r=r_g(\Omega)}^{r_g(\Omega)+H(\Omega)} \varrho(r,\Omega) r^2 dr \tag{2.3}$$

in any direction Ω.

Provided that the density of the condensation layer is chosen in agreement with the principle of mass conservation, the residual topographical potential δV does not include, by definition, zeroth-degree spherical harmonic, but it does include the first-degree spherical harmonics. It also means that the center of mass of the Helmert body, consisting of masses M^g below the geoid plus the mass M^c of the condensation layer, is shifted from the origin of the geocentric coordinate system. Let us compute the magnitude of this shift.

The Cartesian coordinates of the mass-center of the actual Earth's masses M, $M := M^g + M^t$, are defined by formula:

$$\mathbf{x}_T := \frac{1}{M} \int_{\Omega_0} \int_{r=0}^{r_g(\Omega)+H(\Omega)} \varrho(r,\Omega) \mathbf{e}_r(\Omega) r^3 dr d\Omega \ , \tag{2.4}$$

where $\mathbf{e}_r(\Omega)$ is the unit position vector in the direction Ω,

$$\mathbf{e}_r(\Omega) = \sin\vartheta \cos\lambda\, \mathbf{e}_x + \sin\vartheta \sin\lambda\, \mathbf{e}_y + \cos\vartheta\, \mathbf{e}_z \ , \tag{2.5}$$

and the set of unit vectors $\mathbf{e}_x$, $\mathbf{e}_y$, and $\mathbf{e}_z$ forms the Cartesian base vectors. According to the assumption introduced in sect.1.2, the mass-center of the Earth is located at the origin of the coordinate system, i.e.,

$$\mathbf{x}_T = \mathbf{0} \ , \tag{2.6}$$

which means that

$$\int_{\Omega_0} \int_{r=0}^{r_g(\Omega)} \varrho(r,\Omega) \mathbf{e}_r(\Omega) r^3 dr d\Omega + \int_{\Omega_0} \int_{r=r_g(\Omega)}^{r_g(\Omega)+H(\Omega)} \varrho(r,\Omega) \mathbf{e}_r(\Omega) r^3 dr d\Omega = 0 \ . \tag{2.7}$$

Under the assumption that the condensation of the topographical masses conserves the topographical masses, the Cartesian coordinates of the mass-center of the Helmert body of the mass $M^h := M^g + M^c = M$ are

$$\mathbf{x}_T^h = \frac{1}{M} \left[\int_{\Omega_0} \int_{r=0}^{r_g(\Omega)} \varrho(r,\Omega) \mathbf{e}_r(\Omega) r^3 dr d\Omega + \int_{\Omega_0} \sigma(\Omega) \mathbf{e}_r(\Omega) r_g^3(\Omega) d\Omega \right] \ . \tag{2.8}$$

Substituting for the first integral from eqn.(2.7), we get

$$\mathbf{x}_T^h = \frac{1}{M} \left[-\int_{\Omega_0} \int_{r=r_g(\Omega)}^{r_g(\Omega)+H(\Omega)} \varrho(r,\Omega) \mathbf{e}_r(\Omega) r^3 dr d\Omega + \int_{\Omega_0} \sigma(\Omega) \mathbf{e}_r(\Omega) r_g^3(\Omega) d\Omega \right] \ . \tag{2.9}$$

Considering eqn.(2.3) for $\sigma(\Omega)$, we finally get

$$\boxed{\mathbf{x}_T^h = \frac{1}{M} \int_{\Omega_0} \int_{r=r_g(\Omega)}^{r_g(\Omega)+H(\Omega)} \varrho(r,\Omega) \left[r_g(\Omega) - r \right] \mathbf{e}_r(\Omega) r^2 dr d\Omega \ .} \tag{2.10}$$

To estimate the magnitude of the vector $\mathbf{x}_T^h$, we approximate the density $\varrho(r,\Omega)$ of topographical masses by a mean value ϱ_0 and the radius $r_g(\Omega)$ of the geoid by a mean radius R of the Earth. Performing integration over r in eqn.(2.10), we obtain

$$\mathbf{x}_T^h \doteq -\frac{R^2 \varrho_0}{2M} \int_{\Omega_0} H^2(\Omega)\left(1 + \frac{4}{3}\frac{H(\Omega)}{R} + \frac{1}{2}\frac{H^2(\Omega)}{R^2}\right) \mathbf{e}_r(\Omega)d\Omega . \tag{2.11}$$

Moreover, if we take only the first term within the brackets in eqn.(2.11) which is at least of three orders in magnitude larger then the rest, we have

$$\mathbf{x}_T^h \doteq -K \int_{\Omega_0} H^2(\Omega)\mathbf{e}_r(\Omega)d\Omega , \tag{2.12}$$

where

$$K := \frac{R^2 \varrho_0}{2M} \doteq 9 \times 10^{-9}\,\mathrm{m}^{-1} \tag{2.13}$$

for $M = 5.97 \times 10^{24}$ kg, $R = 6371$ km, and $\varrho_0 = 2.67$ g/cm^3. Realizing that

$$\begin{aligned}
\sin\vartheta\cos\lambda &= -\sqrt{\frac{2\pi}{3}}\,[Y_{11}(\Omega) + Y_{11}^*(\Omega)] , \\
\sin\vartheta\sin\lambda &= i\sqrt{\frac{2\pi}{3}}\,[Y_{11}(\Omega) - Y_{11}^*(\Omega)] , \\
\cos\vartheta &= \sqrt{\frac{4\pi}{3}}Y_{10}(\Omega) ,
\end{aligned} \tag{2.14}$$

where $Y_{1m}(\Omega)$ are the first-degree spherical harmonics normalized according to Varshalovich et al. (1989, chapt.5), $i = \sqrt{-1}$, and asterisk denotes complex conjugation, the unit position vector $\mathbf{e}_r(\Omega)$ takes the form

$$\mathbf{e}_r(\Omega) = -\sqrt{\frac{2\pi}{3}}\,[Y_{11}(\Omega) + Y_{11}^*(\Omega)]\,\mathbf{e}_x + i\sqrt{\frac{2\pi}{3}}\,[Y_{11}(\Omega) - Y_{11}^*(\Omega)]\,\mathbf{e}_y + \sqrt{\frac{4\pi}{3}}Y_{10}(\Omega)\,\mathbf{e}_z . \tag{2.15}$$

Substituting eqn.(2.15) into (2.12), introducing spherical harmonic coefficients $(H^2)_{1m}$ as

$$\left(H^2\right)_{1m} := \int_{\Omega_0} H^2(\Omega)Y_{1m}^*(\Omega)d\Omega , \tag{2.16}$$

the Cartesian coordinates (x_T^h, y_T^h, z_T^h) of the vector $\mathbf{x}_T^h$ are

$$\begin{aligned}
x_T^h &= K\sqrt{\frac{2\pi}{3}}\left[\left(H^2\right)_{11} + \left(H^2\right)_{11}^*\right] = 2K\sqrt{\frac{2\pi}{3}}\,Re\left(H^2\right)_{11} , \\
y_T^h &= -K\sqrt{\frac{2\pi}{3}}i\left[-\left(H^2\right)_{11} + \left(H^2\right)_{11}^*\right] = -2K\sqrt{\frac{2\pi}{3}}\,Im\left(H^2\right)_{11} , \\
z_T^h &= -K\sqrt{\frac{4\pi}{3}}\left(H^2\right)_{10} ,
\end{aligned} \tag{2.17}$$

where Re and Im stand for the real and imaginary part of a complex number.

The numerical values of the coefficients $(H^2)_{1m}$ were found for the Earth's topography described by the TUG87 global digital terrain model (Wieser, 1987) complete to degree and order 180. We have obtained

$$\left(H^2\right)_{10} = -0.847 \times 10^5 \, \text{m}^2 \, ,$$
$$\left(H^2\right)_{11} = (-0.207 \times 10^6; \, 0.509 \times 10^6) \, \text{m}^2 \, . \tag{2.18}$$

Then

$$\mathbf{x}_T^h = (-6; \, -15; \, 2) \times 10^{-3} \, \text{m} \, , \tag{2.19}$$

and its magnitude is equal to

$$|\mathbf{x}_T^h| = 16 \times 10^{-3} \, \text{m} \, . \tag{2.20}$$

We can conclude that the center of mass of the Helmert body is shifted about 16 mm from the center of mass of the actual Earth's body in the case when the Helmert condensation is performed according to principle of conservation of topographical masses.

2.2 Principle of mass-center conservation

Now, let us turn to another way of the condensation of topographical masses and assume that they are condensed according to principle of mass-center conservation. This principle postulates that the mass-center of the Helmert body is located at the same position as the mass-center of the actual Earth. In other words, the Cartesian coordinates of the mass-center of the Helmert body are equal to zero in the geocentric coordinate system, i.e.,

$$\mathbf{x}_T^h = \mathbf{0} \, . \tag{2.21}$$

Accounting for eqn.(2.9), the condition (2.21) will be satisfied identically if the condensation density $\sigma(\Omega)$ in an arbitrary direction Ω is, for instance, of the form

$$\sigma(\Omega) = \frac{1}{r_g^3(\Omega)} \int_{r=r_g(\Omega)}^{r_g(\Omega)+H(\Omega)} \varrho(r,\Omega) r^3 dr \, . \tag{2.22}$$

In the case when the condensation density is chosen according to eqn.(2.22), the residual topographical potential δV has no the first-degree spherical harmonics, but it contains the zeroth-degree harmonic. It means that the mass of the condensation layer differs from the topographical masses. The difference between those masses generates the gravitational field that contributes to the geoidal heights. Let us evaluate this contribution.

The mass difference δM between the topographical masses M^t and the mass M^c of the condensation layer may be arranged as follows:

$$\delta M := M^t - M^c = \int_{\Omega_0} \int_{r=r_g(\Omega)}^{r_g(\Omega)+H(\Omega)} \varrho(r,\Omega) r^2 dr d\Omega - \int_{\Omega_0} \sigma(\Omega) r_g^2(\Omega) d\Omega =$$

$$= \int_{\Omega_0} \int_{r=r_g(\Omega)}^{r_g(\Omega)+H(\Omega)} \varrho(r,\Omega) \left[1 - \frac{r}{r_g(\Omega)} \right] r^2 dr d\Omega \ , \qquad (2.23)$$

where we have substituted for masses M^t and M^c from eqn.(2.2), and have used the condensation density (2.22). To estimate the magnitude of δM, we approximately put $\varrho(r,\Omega) \doteq \varrho_0 = 2.67$ g/cm^3, and $r_g(\Omega) \doteq R$. Performing integration over r in eqn.(2.23), we obtain

$$\delta M \doteq -\frac{R\varrho_0}{2} \int_{\Omega_0} H^2(\Omega) \left(1 + \frac{4H(\Omega)}{3R} + \frac{H^2(\Omega)}{2R^2} \right) d\Omega \doteq -\frac{R\varrho_0}{2} \int_{\Omega_0} H^2(\Omega) d\Omega \ , \qquad (2.24)$$

where we have approximated the expression within the brackets by 1.

The correction δN to the geoidal heights due to the mass difference δM is (Heiskanen and Moritz, 1967, eqn.(2-172))

$$\delta N = \frac{G \delta M}{\gamma_Q R} \ , \qquad (2.25)$$

where G is the gravitational constant, and γ_Q is the normal gravity on the level ellipsoid. Substituting for δM from eqn.(2.24), we get

$$\delta N = -\frac{G\varrho_0}{2\gamma_Q} \int_{\Omega_0} H^2(\Omega) d\Omega \ . \qquad (2.26)$$

The averaged value of H^2 over the Earth for the TUG87 global digital terrain model (Wieser, 1987) has been computed numerically. We obtained

$$\int_{\Omega_0} H^2(\Omega) d\Omega = 5.49 \times 10^6 \, \text{m}^2 \ . \qquad (2.27)$$

The correction δN to geoidal heights then takes the value:

$$\delta N = -4.9 \times 10^{-2} \, \text{m} \ . \qquad (2.28)$$

We can conclude that the geoidal heights have to be corrected of about 5 cm in the case that the condensation of the topographical masses is carried out according to the principle mass-center conservation.

2.3   Conclusion

In summary, when topographical masses are condensed on the geoid in the form of a mass layer, the gravitational potential of the Helmert body, consisting of the masses below the geoid plus the mass of the condensation layer, includes the zeroth- and first-degree spherical harmonics. Choosing a suitable prescription for the condensation density, these harmonics may be partly eliminated. Particularly, condensing topographical masses according to the principle of mass conservation, the zeroth-degree spherical harmonic of the anomalous gravitational potential T^h

vanishes, but the first-degree harmonics do not. As a consequence, the center of mass of the Helmert body is shifted from the center of mass of the actual Earth by 1.6 cm approximately.

In other case when the condensation of the topographical masses follows the principle of mass-center conservation, the gravitational center of the condensed masses coincides with that of the actual Earth's masses. By definition, the first-degree spherical harmonics of the potential T^h vanish, but the zeroth-degree harmonic do not. We have shown that this harmonic contributes to the geoidal heights by a constant term of magnitude 5 cm.

Chapter 3

Topographical effects

In the theory introduced in the preceding chapter, the effect of topographical masses on geoid height determination is described by three terms: *the direct topographical effect on gravity* δA, see eqn.(1.32), *the primary indirect topographical effect on potential* δV_{P_g}, see eqn.(1.24), and *the secondary indirect topographical effect on gravity* δS, see eqn.(1.41). All these topographical effects are expressed by means of the residual topographical potential δV which is defined as the difference between the gravitational potential V^t of topographical masses and the gravitational potential V^c of compensating masses, $\delta V = V^t - V^c$, see eqn.(1.10). That is why, we shall firstly express δV by formulae that are suitable for numerical computation, and then, we will examine the three topographical effects separately.

3.1 Approximations used for δV

As shown by Wichiencharoen (1982), Wang and Rapp (1990), or Martinec and Vaníček (1994a), the equipotential surface undulations generated by the residual potential δV for a constant density of 2.67 g/cm^3 are of the order of a metre; they can reach 9 metres for the Mount Everest. For the purpose of computation of the residual potential δV, the geoid will be modelled by a sphere of radius R,

$$r_g(\Omega) = R \, , \tag{3.1}$$

where R is the mean radius of the Earth. This approximation is justifiable because the error introduced by this approximation is at most 3×10^{-3} which then causes an error of at most 27 mm in the geoidal height.

Due to the same reason, the density of topographical masses in the residual potential δV may be modelled by a mean value $\overline{\rho}(\Omega)$ of the actual density $\rho(r, \Omega)$ averaged along the topographical column of height $H(\Omega)$,

$$\overline{\varrho}(\Omega) := \frac{1}{H(\Omega)} \int_{r=R}^{R+H(\Omega)} \varrho(r, \Omega) r^2 dr \, , \tag{3.2}$$

Note that this assumption still allows us to consider lateral density variations in topographical masses which is a more general model than the usual assumption of constant density (Wichiencharoen, 1982; Vaníček and Kleusberg, 1987; Heck, 1993).

Under these assumptions, the topographical potential V^t, eqn.(1.5), and the condensation potential V^σ, eqn.(1.8), takes the following forms:

$$V^t(r, \Omega) = G \int_{\Omega_0} \overline{\varrho}(\Omega') \int_{r'=R}^{R+H(\Omega')} \frac{r'^2}{L(r, \psi, r')} dr' d\Omega' , \qquad (3.3)$$

and

$$V^\sigma(r, \Omega) = GR^2 \int_{\Omega_0} \frac{\sigma(\Omega')}{L(r, \psi, R)} d\Omega' . \qquad (3.4)$$

Moreover, in the following text, we will only use extremely idealized isostatic compensation models with compensation density ϱ_c depending on the angular coordinates Ω only. For such a compensation density, the isostatic compensation potential (1.7) can be simplified as

$$V^\varrho(r, \Omega) = G \int_{\Omega_0} \varrho_c(\Omega') \int_{r'=R-D(\Omega')}^{R} \frac{r'^2}{L(r, \psi, r')} dr' d\Omega' . \qquad (3.5)$$

Under assumptions (3.1) and (3.2), the topographical effects take the following forms: the direct topographical effect on gravity,

$$\delta A(\Omega) = \left. \frac{\partial \, \delta V(r, \Omega)}{\partial r} \right|_{r=R+H(\Omega)} , \qquad (3.6)$$

the primary indirect topographical effect on potential,

$$\delta V(R, \Omega) , \qquad (3.7)$$

and the secondary indirect topographical effect on gravity,

$$\delta S(\Omega) = \frac{2}{R} \delta V(R, \Omega) . \qquad (3.8)$$

The first and the last terms affect the gravitation rather than the potential.

3.2 A weak singularity of the Newton kernel

In this section we remove the singularity of the Newton kernel right in the definition of the Newton integral. By subtracting and adding the value of the Newton kernel at the computation point to the Newton integrand, the singular computation point may be left out from the integration domain. This necessitates the evaluation of Newton's integral over a fixed height and a fixed mass density, which can be done analytically, resulting in the gravitational potential of a spherical Bouguer shell.

As a matter of fact, the Newton kernel $1/L(r, \psi, r')$ grows to infinity when the dummy point (r', Ω') moves towards the computation point (r, Ω). However, the Newton kernel is only **weakly singular** which means that for $r \neq 0$:

$$\lim_{\substack{\psi \to 0 \\ r' \to r}} \frac{\sin \psi}{L(r, \psi, r')} = \frac{1}{r} < \infty . \tag{3.9}$$

Writing the element $d\Omega'$ of the full solid angle in polar coordinates (ψ, α) as $d\Omega' = \sin \psi d\psi d\alpha$, the weak singularity property (3.9) is easily seen reflected also in the corresponding integral form

$$\int_{\Omega_0} \frac{d\Omega'}{L(r, \psi, r')} < \infty . \tag{3.10}$$

This inequality is valid for all non-zero radii r and r'.

The property (3.10) may now be used for removing the weak singularity of the Newton integral (3.3). Subtracting and adding a term

$$V^B(r, \Omega) := G \overline{\varrho}(\Omega) \int_{\Omega_0} \int_{r'=R}^{R+H(\Omega)} \frac{r'^2}{L(r, \psi, r')} dr' d\Omega' \tag{3.11}$$

to the potential $V^t(r, \Omega)$, we get

$$V^t(r, \Omega) = V^B(r, \Omega) + V^R(r, \Omega) , \tag{3.12}$$

where

$$V^R(r, \Omega) := \tag{3.13}$$

$$:= G \int_{\Omega_0} \left[\varrho(\Omega') \int_{r'=R}^{R+H(\Omega')} \frac{r'^2}{L(r, \psi, r')} dr' - \overline{\varrho}(\Omega) \int_{r'=R}^{R+H(\Omega)} \frac{r'^2}{L(r, \psi, r')} dr' \right] d\Omega' .$$

The quantity V^B is easily recognized as the potential of a spherical Bouguer shell of density $\overline{\varrho}(\Omega)$ and thickness $H(\Omega)$. This potential reads (Wichiencharoen, 1982)

$$V^B(r, \Omega) = \begin{cases} 4\pi G \overline{\varrho}(\Omega) \frac{1}{r} \left[R^2 H(\Omega) + R H^2(\Omega) + \frac{1}{3} H^3(\Omega) \right] , & r \geq R + H(\Omega) , \\[2ex] 2\pi G \overline{\varrho}(\Omega) \left[[R + H(\Omega)]^2 - \frac{2}{3} \frac{R^3}{r} - \frac{1}{3} r^2 \right] , & R \leq r \leq R + H(\Omega) , \\[2ex] 4\pi G \overline{\varrho}(\Omega) \left[R H(\Omega) + \frac{1}{2} H^2(\Omega) \right] , & r \leq R . \end{cases} \tag{3.14}$$

Since the actual Earth's surface deviates from the Bouguer sphere (of radius $R + H(\Omega)$), there are deficiencies and/or abundances of topographical masses with respect to the mass of the Bouguer shell. These contribute to the topographical potential $V^t(r, \Omega)$ through the term $V^R(r, \Omega)$ – an analogue of the terrain correction (Heiskanen and Moritz, 1967, sect.3-3.). We will call $V^R(r, \Omega)$ the *terrain roughness term*. It depends chiefly on the behaviour of the difference $H(\Omega) - H(\Omega')$.

Let us now investigate the limit for $\psi \to 0$ of the subintegral function in the angular integral (3.13). When $\psi \to 0$, then $\bar{\varrho}(\Omega') \to \bar{\varrho}(\Omega)$ and $H(\Omega') \to H(\Omega)$. We will assume, reasonably, that both the topographical density $\bar{\varrho}$ and the topographical height H are bounded (i.e., that there are no mass-singularities inside the topographical masses and the heights of the Earth's topography are finite). Then the limit for $\psi \to 0$ of the subintegral function in eqn.(3.13) reads

$$\lim_{\psi \to 0} \left[\bar{\varrho}(\Omega') \int_{r'=R}^{R+H(\Omega')} \frac{r'^2}{L(r, \psi, r')} dr' - \bar{\varrho}(\Omega) \int_{r'=R}^{R+H(\Omega)} \frac{r'^2}{L(r, \psi, r')} dr' \right] \sin \psi =$$

$$= \bar{\varrho}(\Omega) \int_{r'=R}^{R+H(\Omega)} \lim_{\psi \to 0} \frac{\sin \psi}{L(r, \psi, r')} r'^2 dr' - \bar{\varrho}(\Omega) \int_{r'=R}^{R+H(\Omega)} \lim_{\psi \to 0} \frac{\sin \psi}{L(r, \psi, r')} r'^2 dr' . \quad (3.15)$$

Since both functions $\bar{\varrho}(\Omega)$ and $H(\Omega)$ are bounded and the Newton kernel is weakly singular, see property (3.9), both integrals on the right-hand side of eqn.(3.15) are finite and have the same value; their difference is thus equal to zero. This means that the point $\psi = 0$ can be removed from the integration domain Ω_0 of the integral (3.13). This fact is important for the numerical computation of the topographical potential $V^t(r, \Omega)$ because the formulae (3.12) and (3.13) guarantee that the numerical algorithm is not forced to evaluate the undefined expression of the type of $0/0$ encountered in the original Newton integral (3.3).

3.3 The Pratt-Hayford and the Airy-Heiskanen isostatic compensation models

In order to remove the singularity of the isostatic compensation potential $V^\varrho(r, \Omega)$, cf. eqn.(3.5), we may proceed in a way analogous to that for the potential V^t:

$$V^\varrho(r, \Omega) = V^{\varrho,B}(r, \Omega) + V^{\varrho,R}(r, \Omega) , \quad (3.16)$$

where the Bouguer term is equal to

$$V^{\varrho,B}(r, \Omega) := G\varrho_c(\Omega) \int_{\Omega_0} \int_{r'=R-D(\Omega')}^{R} \frac{r'^2}{L(r, \psi, r')} dr' d\Omega' =$$

$$= 4\pi G\varrho_c(\Omega) \frac{1}{r} \left[R^2 D(\Omega) - R D^2(\Omega) + \frac{1}{3} D^3(\Omega) \right] \quad \text{if } r \geq R , \quad (3.17)$$

and the terrain roughness term reads

$$V^{\varrho,R}(r, \Omega) :=$$

$$:= G \int_{\Omega_0} \left[\varrho_c(\Omega') \int_{r'=R-D(\Omega')}^{R} \frac{r'^2}{L(r, \psi, r')} dr' - \varrho_c(\Omega) \int_{r'=R-D(\Omega)}^{R} \frac{r'^2}{L(r, \psi, r')} dr' \right] d\Omega' . \quad (3.18)$$

The compensation density $\varrho_c(\Omega)$ and the depth $D(\Omega)$ are two free parameters of an isostatic compensation model that must be chosen before computing topographical effects. We will use the simplest mechanism of compensation of topographical masses by assuming that the compensation is **strictly local**. This means that, at a point on the Earth's surface, the gravitational effect of the Bouguer term $V^B(r, \Omega)$ of topographical potential V^t is compensated by the gravitational effect of the Bouguer term $V^{\varrho,B}(r, \Omega)$ of isostatic potential V^ϱ, i.e.,

$$V^B(r, \Omega) = V^{\varrho,B}(r, \Omega) \qquad \text{for } r = R + H(\Omega) . \tag{3.19}$$

Substituting eqns.(3.14) and (3.17) into the last constrain, we get

$$\overline{\varrho}(\Omega)H(\Omega)\left[1 + \frac{H(\Omega)}{R} + \frac{H^2(\Omega)}{3R^2}\right] = \varrho_c(\Omega)D(\Omega)\left[1 - \frac{D(\Omega)}{R} + \frac{D^2(\Omega)}{3R^2}\right] . \tag{3.20}$$

We can see that the two parameters $\varrho_c(\Omega)$ and $D(\Omega)$ are constrained by one condition only. Hence, the assumption that the compensation of topographical masses has a strictly local character does not yield a unique determination of $\varrho_c(\Omega)$ and $D(\Omega)$, and an additional information must be supplied to eqn.(3.20) to determine $\varrho_c(\Omega)$ and $D(\Omega)$ uniquely.

In the past, two extremely idealized isostatic compensation models were proposed to cancel the effect of topographical abundances from surface gravity observations. In *the Pratt-Hayford model* (e.g., Heiskanen and Moritz, 1967, sect.3-4.), the topographical masses are compensated by varying density distribution within the layer of a constant thickness, $D(\Omega) = D_0$=const. The compensation density $\varrho_c(\Omega)$ is then given by eqn.(3.20) as

$$\varrho_c(\Omega) = \overline{\varrho}(\Omega)\frac{H(\Omega)}{D_0}\frac{1 + \dfrac{H(\Omega)}{R} + \dfrac{H^2(\Omega)}{3R^2}}{1 - \dfrac{D_0}{R} + \dfrac{D_0^2}{3R^2}} . \tag{3.21}$$

Since $H(\Omega)/R \ll 1$ and $D_0/R \ll 1$ (as usual, we will put $D_0 = 100$ km (Heiskanen and Moritz, 1967, eqn.(3-29))), the last fraction in eqn.(3.21) can be approximated by 1. Moreover, the Pratt-Hayford model considers the density $\overline{\varrho}(\Omega)$ of topographical masses constant equal to the mean crustal density $\varrho_0 = 2.67$ g/cm^3. Eqn.(3.21) can then be simplified as

$$\boxed{\varrho_c(\Omega) \doteq \varrho_0\frac{H(\Omega)}{D_0}} . \tag{3.22}$$

By the Pratt-Hayford compensation density (3.22), the terrain roughness term (3.18) reads

$$\boxed{V^{\varrho,R}(r, \Omega) = G\frac{\varrho_0}{D_0}\int_{\Omega_0} [H(\Omega') - H(\Omega)]\int_{r'=R-D_0}^{R} \frac{r'^2}{L(r, \psi, r')}dr' d\Omega'} . \tag{3.23}$$

The Airy-Heiskanen model assumes that the topographical masses are compensated by varying thickness $D(\Omega)$ of a compensation layer. The density of compensation layer is considered constant equal to the density contrast $\Delta\varrho_{Moho}$ (=const > 0) at the Moho discontinuity, i.e., $\varrho_c(\Omega) = \Delta\varrho_{Moho}$. Eqn.(3.20) then represents the cubic equation for compensation depth $D(\Omega)$. Considering again that $H(\Omega)/R \ll 1$ and $D(\Omega)/R \ll 1$, and putting $\bar{\varrho}(\Omega) = \varrho_0$, the solution to eqn.(3.20) may approximately be written as

$$D(\Omega) \doteq H(\Omega)\frac{\varrho_0}{\Delta\varrho_{Moho}} \ . \tag{3.24}$$

The density jump $\Delta\varrho_{Moho}$ will be taken according to Martinec (1994a), $\Delta\varrho_{Moho} = 0.28$ g/cm^3.

Finally, the terrain roughness term (3.18) for the Airy-Heiskanen compensation model may be simplified as

$$V^{\varrho,R}(r,\Omega) = G\Delta\varrho_{Moho} \int_{\Omega_0} \int_{r'=R-D(\Omega')}^{R-D(\Omega)} \frac{r'^2}{L(r,\psi,r')}dr'd\Omega' \ . \tag{3.25}$$

3.4 Helmert's condensation layer

The potential (3.4) of Helmert's condensation layer may be expressed analogously to eqn.(3.12),

$$V^\sigma(r,\Omega) = V^{\sigma,B}(r,\Omega) + GR^2 \int_{\Omega_0} \frac{\sigma(\Omega') - \sigma(\Omega)}{L(r,\psi,R)}d\Omega' \ . \tag{3.26}$$

Here the symbol $V^{\sigma,B}(r,\Omega)$ denotes the gravitational potential of a spherical layer with density $\sigma(\Omega)$ and radius R:

$$V^{\sigma,B}(r,\Omega) := GR^2\sigma(\Omega) \int_{\Omega_0} \frac{d\Omega'}{L(r,\psi,R)} \ . \tag{3.27}$$

The last integral may be readily evaluated yielding

$$V^{\sigma,B}(r,\Omega) = \begin{cases} 4\pi G\sigma(\Omega)\dfrac{R^2}{r} \ , & r > R \ , \\[2ex] 4\pi G\sigma(\Omega)R \ , & r \leq R \ . \end{cases} \tag{3.28}$$

The 2-D condensation density $\sigma(\Omega)$ can be chosen in a variety of ways. Here, we will again follow the principle of strictly local compensation, i.e., we will assume that

$$V^B(r,\Omega) = V^{\sigma,B}(r,\Omega) \qquad \text{for} \ \ r = R + H(\Omega) \ , \tag{3.29}$$

where $V^B(r, \Omega)$ is the Bouguer term of topographical potential V^t. Substituting eqns.(3.14) and (3.28) into (3.29), the condensation density $\sigma(\Omega)$ reads

$$\boxed{\sigma(\Omega) = \bar{\varrho}(\Omega)H(\Omega)\left(1 + \frac{H(\Omega)}{R} + \frac{H^2(\Omega)}{3R^2}\right) \,.} \tag{3.30}$$

Note that this density of the Helmert condensation layer is identical with the condensation density (2.3) chosen according to principle of mass conservation when the approximations (3.1) and (3.2) are used.

3.5   The direct topographical effect on gravity

Now, we are ready to investigate the topographical effects defined by eqns.(3.6)–(3.8). Let us start with the direct topographical effect on gravity $\delta A(\Omega)$. This effect is nothing but the radial derivative of the residual potential δV taken at a point $(R + H(\Omega), \Omega)$ on the Earth's surface, cf., eqn.(3.6). Substituting for the residual topographical potential δV from eqn.(1.10) into eqn.(3.6), we can write

$$\delta A(\Omega) = A^t(\Omega) - A^c(\Omega) \,, \tag{3.31}$$

where

$$A^t(\Omega) := \frac{\partial V^t(r, \Omega)}{\partial r}\bigg|_{r=R+H(\Omega)} \,, \qquad A^c(\Omega) := \frac{\partial V^c(r, \Omega)}{\partial r}\bigg|_{r=R+H(\Omega)} \,, \tag{3.32}$$

are the radial components of the gravitational attraction induced by the topographical and compensated masses at the point on the Earth's surface.

Taking the radial derivative of eqn.(3.12), and putting $r = R + H(\Omega)$, we obtain

$$A^t(\Omega) = A^B(\Omega) +$$

$$+G\int_{\Omega_0}\left[\bar{\varrho}(\Omega')\frac{\partial \widetilde{L^{-1}}(r, \psi, r')}{\partial r}\bigg|_{r'=R}^{R+H(\Omega')} - \bar{\varrho}(\Omega)\frac{\partial \widetilde{L^{-1}}(r, \psi, r')}{\partial r}\bigg|_{r'=R}^{R+H(\Omega)}\right]_{r=R+H(\Omega)} d\Omega' \,, \tag{3.33}$$

where $A^B(\Omega)$ is the radial component of the attraction of the spherical Bouguer shell at the point $(R + H(\Omega), \Omega)$, i.e.,

$$A^B(\Omega) := \frac{\partial V^B(r, \Omega)}{\partial r}\bigg|_{r=R+H(\Omega)} \,. \tag{3.34}$$

To abbreviate notations, we have introduced symbol $\widetilde{L^{-1}}(r, \psi, r')$ for an indefinite radial integral of the Newton kernel,

$$\widetilde{L^{-1}}(r, \psi, r') := \int_{r'}\frac{r'^2}{L(r, \psi, r')}dr' \,. \tag{3.35}$$

Provided that the topographical masses are compensated isostatically, i.e., when $V^c = V^\varrho$, the attraction of compensated masses can be derived taking the radial derivatives of eqn.(3.16). We immediately get

$$A^c(\Omega) = A^{\varrho,B}(\Omega)+ \tag{3.36}$$

$$+G \int_{\Omega_0} \left[\varrho_c(\Omega') \frac{\partial \widetilde{L^{-1}}(r,\psi,r')}{\partial r} \bigg|_{r'=R-D(\Omega')}^{R} - \varrho_c(\Omega) \frac{\partial \widetilde{L^{-1}}(r,\psi,r')}{\partial r} \bigg|_{r'=R-D(\Omega)}^{R} \right]_{r=R+H(\Omega)} d\Omega',$$

where

$$A^{\varrho,B}(\Omega) := \frac{\partial V^{\varrho,B}(r,\Omega)}{\partial r} \bigg|_{r=R+H(\Omega)} . \tag{3.37}$$

Moreover, when the isostatic compensation is strictly local, see constrain (3.19), then

$$A^B(\Omega) = A^{\varrho,B}(\Omega) . \tag{3.38}$$

As a consequence, the direct topographical effect on gravity $\delta A(\Omega)$ is composed from the terrain roughness terms only, i.e.,

$$\delta A(\Omega) =$$

$$= G \int_{\Omega_0} \left[\overline{\varrho}(\Omega') \frac{\partial \widetilde{L^{-1}}(r,\psi,r')}{\partial r} \bigg|_{r'=R}^{R+H(\Omega')} - \overline{\varrho}(\Omega) \frac{\partial \widetilde{L^{-1}}(r,\psi,r')}{\partial r} \bigg|_{r'=R}^{R+H(\Omega)} \right]_{r=R+H(\Omega)} d\Omega' - \tag{3.39}$$

$$-G \int_{\Omega_0} \left[\varrho_c(\Omega') \frac{\partial \widetilde{L^{-1}}(r,\psi,r')}{\partial r} \bigg|_{r'=R-D(\Omega')}^{R} - \varrho_c(\Omega) \frac{\partial \widetilde{L^{-1}}(r,\psi,r')}{\partial r} \bigg|_{r'=R-D(\Omega)}^{R} \right]_{r=R+H(\Omega)} d\Omega'.$$

Particularly, if the isostatic compensation is governed by the <u>Pratt-Hayford</u> <u>model</u>, c.f., sect.3.3, and the density of topographical masses is constant, $\overline{\varrho}(\Omega) = \varrho_0$, the direct topographical effect on gravity can be simplified as follows,

$$\delta A(\Omega) = G\varrho_0 \int_{\Omega_0} \left[\frac{\partial \widetilde{L^{-1}}(r,\psi,r')}{\partial r} \bigg|_{r'=R+H(\Omega)}^{R+H(\Omega')} - \tag{3.40} \right.$$

$$\left. - \frac{H(\Omega') - H(\Omega)}{D_0} \frac{\partial \widetilde{L^{-1}}(r,\psi,r')}{\partial r} \bigg|_{r'=R-D_0}^{R} \right]_{r=R+H(\Omega)} d\Omega' .$$

For the <u>Airy-Heiskanen model</u> model of the compensation of topographical masses, the direct topographical effect on gravity (3.39) reads

$$\delta A(\Omega) = G\varrho_0 \int_{\Omega_0} \left[\left. \frac{\partial \widetilde{L^{-1}}(r, \psi, r')}{\partial r} \right|_{r'=R+H(\Omega)}^{R+H(\Omega')} \right. \qquad (3.41)$$
$$\left. - \frac{\Delta \varrho_{Moho}}{\varrho_0} \left. \frac{\partial \widetilde{L^{-1}}(r, \psi, r')}{\partial r} \right|_{r'=R-D(\Omega')}^{R-D(\Omega)} \right]_{r=R+H(\Omega)} d\Omega' .$$

Finally, if the <u>Helmert 2nd condensation</u> is used to compensate the gravitational effect of topographical masses, then $V^c = V^\sigma$, and the attraction of the condensed masses can be derived taking the radial derivative of eqn.(3.26). We obtain

$$A^c(\Omega) = A^{\sigma,B}(\Omega) + GR^2 \int_{\Omega_0} [\sigma(\Omega') - \sigma(\Omega)] \left. \frac{\partial L^{-1}(r, \psi, R)}{\partial r} \right|_{r=R+H(\Omega)} d\Omega' , \quad (3.42)$$

where $A^{\sigma,B}(\Omega)$ is the attraction of a spherical single layer with the density $\sigma(\Omega)$:

$$A^{\sigma,B}(\Omega) := \left. \frac{\partial V^\sigma(r, \Omega)}{\partial r} \right|_{r=R+H(\Omega)} . \qquad (3.43)$$

Moreover, assuming that the condensation of topographical masses is strictly local, see constrain (3.29), we get

$$A^B(\Omega) = A^{\sigma,B}(\Omega) . \qquad (3.44)$$

The consequence is that the direct topographical effect $\delta A(\Omega)$ consists of the terrain roughness contributions only:

$$\delta A(\Omega) = G \int_{\Omega_0} \left[\bar{\varrho}(\Omega') \left. \frac{\partial \widetilde{L^{-1}}(r, \psi, r')}{\partial r} \right|_{r'=R}^{R+H(\Omega')} - \bar{\varrho}(\Omega) \left. \frac{\partial \widetilde{L^{-1}}(r, \psi, r')}{\partial r} \right|_{r'=R}^{R+H(\Omega)} \right.$$
$$\left. - R^2 [\sigma(\Omega') - \sigma(\Omega)] \frac{\partial L^{-1}(r, \psi, R)}{\partial r} \right]_{r=R+H(\Omega)} d\Omega' . \qquad (3.45)$$

3.6 The primary indirect topographical effect on potential

To find the expression for the primary indirect topographical effect on potential, the residual topographical potential $\delta V = V^t - V^c$ must be evaluated on the geoid $(r = R)$. The formulae derived in sections 3.2-3.4 may be used directly for determining $\delta V(R, \Omega)$. Eqns.(3.12)–(3.14) yield

$$V^t(R, \Omega) = 4\pi G\bar{\varrho}(\Omega) \left(RH(\Omega) + \frac{1}{2}H^2(\Omega) \right) +$$

$$+G \int_{\Omega_0} \left[\bar{\varrho}(\Omega') \; \widetilde{L^{-1}}(R, \psi, r') \Big|_{r'=R}^{R+H(\Omega')} - \bar{\varrho}(\Omega) \; \widetilde{L^{-1}}(R, \psi, r') \Big|_{r'=R}^{R+H(\Omega)} \right] d\Omega', \quad (3.46)$$

where for the gravitational potential of the spherical Bouguer shell we have substituted from the last of eqn.(3.14). Similarly, by eqns.(3.16)–(3.18), we get

$$V^\varrho(R, \Omega) = 4\pi G\varrho_c(\Omega) \left(RD(\Omega) - D^2(\Omega) + \frac{D^3(\Omega)}{3R} \right) +$$

$$+G \int_{\Omega_0} \left[\varrho_c(\Omega') \; \widetilde{L^{-1}}(R, \psi, r') \Big|_{r'=R-D(\Omega')}^{R} - \varrho_c(\Omega) \; \widetilde{L^{-1}}(R, \psi, r') \Big|_{r'=R-D(\Omega)}^{R} \right] d\Omega'. \quad (3.47)$$

Finally,

$$V^\sigma(R, \Omega) = 4\pi GR\sigma(\Omega) + GR^2 \int_{\Omega_0} \frac{\sigma(\Omega') - \sigma(\Omega)}{L(R, \psi, R)} d\Omega', \quad (3.48)$$

where the last of eqn.(3.28) has been used for the potential of spherical material layer.

In particular, taking the compensation density $\varrho_c(\Omega)$ as defined by eqn.(3.22), the primary indirect topographical effect on potential for the Pratt-Hayford model becomes

$$\delta V(R, \Omega) = 4\pi G\varrho_0 H(\Omega) \left(\frac{1}{2}H(\Omega) + D_0 \right) + \qquad (3.49)$$

$$+G\varrho_0 \int_{\Omega_0} \left[\widetilde{L^{-1}}(R, \psi, r') \Big|_{r'=R+H(\Omega)}^{R+H(\Omega')} - \frac{H(\Omega') - H(\Omega)}{D_0} \; \widetilde{L^{-1}}(R, \psi, r') \Big|_{r'=R-D_0}^{R} \right] d\Omega'.$$

For the Airy-Heiskanen model with the compensation density (3.24), the primary indirect topographical effect on potential reads

$$\delta V(R, \Omega) = 4\pi G\varrho_0 H^2(\Omega) \left(\frac{1}{2} + \frac{\varrho_0}{\Delta\varrho_{Moho}} \right) + \qquad (3.50)$$

$$+G\varrho_0 \int_{\Omega_0} \left[\widetilde{L^{-1}}(R, \psi, r') \Big|_{r'=R+H(\Omega)}^{R+H(\Omega')} - \frac{\Delta\varrho_{Moho}}{\varrho_0} \; \widetilde{L^{-1}}(R, \psi, r') \Big|_{r'=R-D(\Omega')}^{R-D(\Omega)} \right] d\Omega'.$$

Finally, the primary indirect topographical effect on potential for Helmert's condensation layer becomes

$$\delta V(R, \Omega) = -2\pi G\overline{\varrho}(\Omega)H^2(\Omega)\left(1 + \frac{2}{3}\frac{H(\Omega)}{R}\right) +$$

$$+ G\int_{\Omega_0}\left[\overline{\varrho}(\Omega')\left.\widetilde{L^{-1}}(R, \psi, r')\right|_{r'=R}^{R+H(\Omega')} - \overline{\varrho}(\Omega)\left.\widetilde{L^{-1}}(R, \psi, r')\right|_{r'=R}^{R+H(\Omega)} - \qquad (3.51)$$

$$- R^2\frac{\sigma(\Omega') - \sigma(\Omega)}{L(R, \psi, R)}\right]d\Omega' .$$

3.7   The secondary indirect topographical effect on gravity

By definition (cf., eqn.(3.8)), the secondary indirect topographical effect on gravity, $\delta S(\Omega)$, is given by the residual topographical potential δV at a point on the geoid multiplied by $2/R$. Hence, this effect may easily be expressed by means of the primary indirect effect on potential $\delta V(R, \Omega)$ for a respective type of the compensation of topographical masses, c.f., formulae (3.49)–(3.51).

3.8   Analytical expressions for integration kernels of Newton's type

An important fact is that an indefinite radial integral of the Newton kernel, i.e., the expression $\widetilde{L^{-1}}(r, \psi, r')$ may be evaluated analytically (Gradshteyn and Ryzhik, 1980, pars. 2.261, 2.264):

$$\widetilde{L^{-1}}(r, \psi, r') = \frac{1}{2}(r' + 3r\cos\psi)L(r, \psi, r') +$$

$$+ \frac{r^2}{2}(3\cos^2\psi - 1)\ln|r' - r\cos\psi + L(r, \psi, r')| + C , \qquad (3.52)$$

where the 'constant' C may depend on the variables r and ψ only.

For computing the direct topographical effect on gravity δA, see sect.3.5, we need to compute the radial derivative of the Newton surface and volume integrals. This requires to determine the radial derivative of the Newton kernel $1/L(r, \psi, r')$ and the radial derivative of the kernel $\widetilde{L^{-1}}(r, \psi, r')$. To find the former derivative is straightforward; by eqn.(1.6), we readily get

$$\frac{\partial L^{-1}(r, \psi, r')}{\partial r} = -\frac{r - r'\cos\psi}{\left(r^2 + r'^2 - 2rr'\cos\psi\right)^{3/2}} . \qquad (3.53)$$

Using eqn.(3.52), we can find an analytical formula for the radial derivative of the kernel $\widetilde{L^{-1}}(r, \psi, r')$. After some algebra, we get

$$\frac{\partial \widetilde{L^{-1}}(r, \psi, r')}{\partial r} = \left[(r'^2 + 3r^2) \cos \psi + (1 - 6 \cos^2 \psi)rr'\right] L^{-1}(r, \psi, r') +$$

$$+ r(3 \cos^2 \psi - 1) \ln |r' - r \cos \psi + L(r, \psi, r')| . \tag{3.54}$$

3.8.1 The singularity of the kernel $\widetilde{L^{-1}}(r, \psi, r')$ at the point $\psi = 0$

Analysing eqn.(3.13), we have learned that there is no need to evaluate the vertically integrated kernel $\widetilde{L^{-1}}(R, \psi, r')\Big|_{r'=R}^{R+H(\Omega')}$ at the computation point ($\psi = 0$). Nevertheless, the kernel has to be evaluated in the immediate neighbourhood of the computation point; therefore, we have to investigate the type of singularity of the kernel $\widetilde{L^{-1}}(R, \psi, r')\Big|_{r'=R}^{R+H(\Omega')}$ at that point. Thus, let us have a look at the behaviour of that kernel in the vicinity of the point $\psi = 0$. Using eqn.(3.52), we obtain

$$\widetilde{L^{-1}}(R, \psi, r')\Big|_{r'=R}^{R+H(\Omega')} =$$

$$= \frac{1}{2}(R + H(\Omega') + 3R \cos \psi)L(R, \psi, R + H(\Omega')) - \frac{1}{2}(R + 3R \cos \psi)\ell_0 +$$

$$+ \frac{R^2}{2}(3 \cos^2 \psi - 1) \ln \frac{H(\Omega') + \frac{\ell_0^2}{2R} + L(R, \psi, R + H(\Omega'))}{\ell_0 + \frac{\ell_0^2}{2R}} , \tag{3.55}$$

where

$$L(R, \psi, R + H(\Omega')) = \sqrt{(R + H(\Omega'))^2 + R^2 - 2R(R + H(\Omega')) \cos \psi} =$$

$$= \sqrt{\ell_0^2 + \frac{H(\Omega')}{R}\ell_0^2 + H^2(\Omega')} , \tag{3.56}$$

and ℓ_0 is the spatial distance between points (R, Ω) and (R, Ω'), i.e.,

$$\ell_0 := L(R, \psi, R) = 2R \sin \frac{\psi}{2} . \tag{3.57}$$

When a dummy point of integration comes close to the computation point ($\psi \to 0$ or $\ell_0 \to 0$), the first two term on the right-hand side of eqn.(3.55) come to a finite number $(4R + H(\Omega'))H(\Omega')/2$; the magnitude of the last term grows to infinity since this term behaves like a function $\ln \ell_0$ (we assume that $H(\Omega') > 0$). Because the following limit is valid

$$\lim_{\ell_0 \to 0} \frac{\ln \ell_0}{\frac{1}{\ell_0}} = 0 , \tag{3.58}$$

the magnitude of the kernel $\widetilde{L^{-1}}(R, \psi, r')\Big|_{r'=R}^{R+H(\Omega')}$ grows to infinity even slower than a reciprocal distance $1/\ell_0$ when ψ approaches zero. Therefore, the numerical procedure of computing the Newton integral (3.13), based on eqn.(3.52) is very stable even near the computation point ($\psi = 0$).

3.9   Numerical tests

For computing the primary indirect topographical effect on potential, we have created the computation code for carrying out the integration kernel

$$\widetilde{L^{-1}}(r,\psi,r')\Big|_{r'=R}^{R+H(\Omega')} \tag{3.59}$$

generating the gravitational potential $V^t(r,\Omega)$ of the topographical masses, the integration kernel

$$R^2\frac{\tau(\Omega')}{L(r,\psi,R)} \tag{3.60}$$

generating the potential $V^\sigma(r,\Omega)$ of Helmert's condensation layer, and their difference, i.e., the kernel

$$\widetilde{L^{-1}}(r,\psi,r')\Big|_{r'=R}^{R+H(\Omega')} - R^2\frac{\tau(\Omega')}{L(r,\psi,R)} \tag{3.61}$$

generating the residual topographical potential $\delta V(r,\Omega)$ for Helmert's 2nd condensation. For the sake of brevity, we have introduced $\tau(\Omega)$ for the expression

$$\tau(\Omega) := H(\Omega)\left(1 + \frac{H(\Omega)}{R} + \frac{H^2(\Omega)}{3R^2}\right) \ . \tag{3.62}$$

Note that the integration kernels of the residual topographical potential $\delta V(r,\Omega)$ for the Pratt-Hayford or Airy-Heiskanen compensation models are easily to set up by means of the kernel (3.59).

Analogously, for computing the direct topographical effect on gravity, we have created the computation code for carrying out the integration kernel

$$\frac{\partial\widetilde{L^{-1}}(r,\psi,r')}{\partial r}\Bigg|_{r'=R}^{R+H(\Omega')} \tag{3.63}$$

generating the radial component of the gravitational attraction of topographical masses, the integration kernel

$$R^2\tau(\Omega)\frac{\partial L^{-1}(r,\psi,R)}{\partial r} \tag{3.64}$$

generating the radial component of the gravitational attraction of Helmert's condensation layer, and their difference, i.e., the kernel

$$\frac{\partial\widetilde{L^{-1}}(r,\psi,r')}{\partial r}\Bigg|_{r'=R}^{R+H(\Omega')} - R^2\tau(\Omega)\frac{\partial L^{-1}(r,\psi,R)}{\partial r} \tag{3.65}$$

generating the direct topographical effect on gravity.

Table 3.1: The minimum and maximum values of the direct and indirect topographical effects over the Canadian Rocky Mountains for various integration radii ψ_0

ψ_0 (degrees)	min mGal	max mGal	min (cm)	max (cm)
0.5	-107.8	47.1	-2.7	4.2
1.0	-117.7	47.2	-2.7	4.9
1.5	-121.1	47.2	-2.6	5.2
2.0	-121.5	47.2	-2.0	5.9
3.0	-121.7	47.2	-2.0	6.7
4.0	-121.8	47.2	-2.1	6.8

We have carried out several numerical tests to explore the behaviour of the above integration kernels in dependence of varying topographical height $H(\Omega')$ and the angular distance ψ. Figure 3.1 summarizes the results of these tests. The angular distance ψ was changed fluently from $0.0001°$ to $1°$, i.e., the distance between the computation point and an integration point ranges approximately from 11 m to 110 km. For the indirect topographical effect on potential (the left column of plots in Figure 3.1) height $H(\Omega')$ of the integration point has been modelled by three values, $H(\Omega') = 200$ m, 100 m, and 5000 m. We have plotted separately the particular parts of the integration kernel (3.61); the solid curves show the behaviour of the kernel (3.59), the dotted curves stand for the kernel (3.60), and the dashed lines show their difference, i.e., the kernel (3.61).

In modelling the properties of kernels (3.63)–(3.65) occurring in the direct topographical effect on gravity, we have fixed the height $H(\Omega')$ of an integration point, $H(\Omega') = 1000$ m, and have varied the height $H(\Omega)$ of the computation point, $H(\Omega) = 200$ m, 1000 m, and 5000 m. The results are plotted in the right column of Figure 3.1. The solid curves show the behaviour of the kernel (3.63), the dotted curves stand for the kernel (3.64), and the dashed lines show their difference, i.e., the kernel (3.65).

Inspecting Figure 3.1, we can observe that

- The computation algorithm for carrying out the above integration kernels is very stable even if a dummy point of integration is very close to the computational point; this confirms our theoretical consideration in sect.3.8.1.

- There is a large difference in magnitudes of kernels (3.59) and (3.60) in an immediate neighbourhood of the computation point. This difference depends on a height of the computation point; the larger the height of the computation point, the larger the difference between these kernels, and therefore, the stronger is the indirect topographical effect on potential. An increase of integration kernel (3.61) for decreasing ψ's also means that the indirect topographical effect will be stronger for a digital terrain model with a tiny

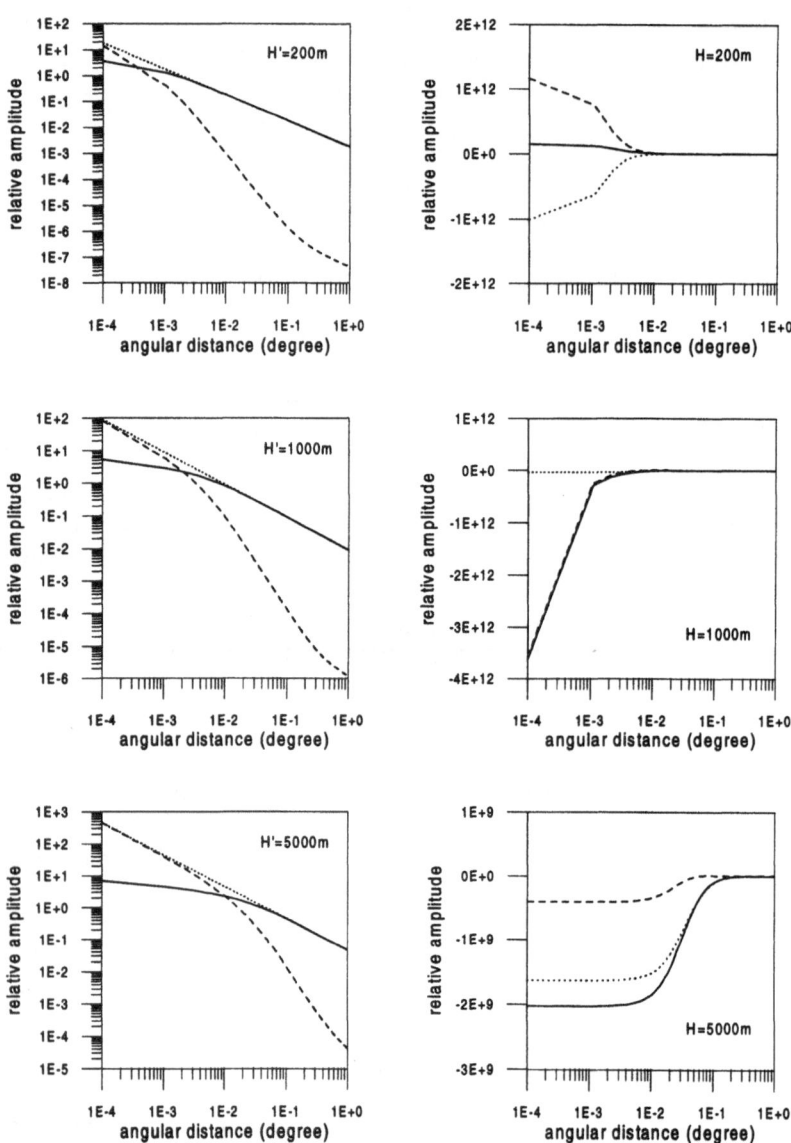

Figure 3.1: The integration kernels for the indirect and direct
topographical effects

grid step size rather than that with a sparse grid size. In sect.5.3.2 we will
show, for instance, that densifying a $5' \times 5'$ grid of topographical heights in
the Canadian Rocky Mountains to a 1 km×1 km grid causes an increase of
the maximal amplitude of the indirect topographical effect from 3 cm to 18
cm.

• When the computation point moves away from the integration point, the integration kernel (3.60) approaches the integration kernel (3.59). For instance, the difference between kernels (3.59) and (3.60) falls 7 orders in magnitude when integration point moves from $\psi = 0.0001°$ to $\psi = 1°$. This confirms the well known-fact that the gravitational potential of topographical masses of a finite thickness behaves like the potential of a thin layer when it is observed from a larger distance. The integration over the full solid angle Ω' may be thus restricted to a small area (of radius ψ_0) surrounding the computation point. A question arises how large the integration radius ψ_0 should be chosen in order to keep a prescribed accuracy of topographical effects. To answer this question, we have computed the direct and indirect topographical effects over the Canadian Rocky Mountains with a $5' \times 5'$ digital terrain model for various radii ψ_0. The results are summarized in Table 3.1 showing that to reach the accuracy of the order of 1 cm in computing the indirect topographical effect on the potential or to reach the accuracy of the order of 0.5 mgal in computing the direct topographical effect on gravity it is sufficient to integrate over a spherical cap of radius of $3°$.

Chapter 4

Planar approximation

The general formulae (3.45) and (3.51) are accurate to a spherical approximation and use the mean radial topographical density (see sect.3.1). They can be further simplified by accepting certain approximations permissible from the accuracy point of view. To be able to compare our results with those derived by Heck (1993), the Helmert condensation technique will only be considered to reduce the gravitational effect of topographical masses; provided that another model of compensation is employed, the approximations presented below can be drawn by an analogous way.

4.1   Constant density of topographical masses

The first approximation is based on the fact that the density of topographical masses varies by about 10% to 20% around the mean crustal density $\varrho_0 = 2.67 \mathrm{g/cm^3}$. In Chapter 6 we will see that these variations contribute at most by several decimetres to the geoidal heights even in the highest mountains. In the terrain with heights up to 1000 m, the contributions to geoidal heights by lateral variations of topographical density are below 1 cm and may be safely neglected. For such a case, the density of topographical masses is modelled by the mean crustal density ϱ_0.

Taking this approximation into account, i.e., putting $\overline{\varrho}(\Omega) = \varrho_0$, the general formula (3.51) for the indirect topographical effect on potential reduces to

$$\delta V(R, \Omega) = -2\pi G\varrho_0 H^2(\Omega) \left(1 + \frac{2}{3}\frac{H(\Omega)}{R}\right) +$$

$$+ G\varrho_0 \int_{\Omega_0} \left[\left.\widetilde{L^{-1}}(R, \psi, r')\right|_{r'=R+H(\Omega)}^{R+H(\Omega')} - R^2 \frac{\tau(\Omega') - \tau(\Omega)}{\ell_0} \right] d\Omega' , \qquad (4.1)$$

where ℓ_0 and $\tau(\Omega)$ are defined by eqn.(3.57) and (3.62), respectively.

4.2 Restricted integration

In sect.3.9 we have demonstrated that the kernel (3.60) approaches the kernel (3.59) when the integration point moves away from the computation point. The integration over angular coordinates Ω' in eqn.(4.1) may be thus limited to a small area (of radius ψ_0) surrounding the computation point. To get a 1 cm accuracy of the residual topographical potential $\delta V(R, \Omega)$, it is sufficient to integrate up to a distance $\psi_0 = 3°$ (see Table 3.1). The formula (4.1) for the indirect topographical effect on potential then becomes

$$\delta V(R, \Omega) \doteq -2\pi G \varrho_0 H^2(\Omega) \left(1 + \frac{2}{3}\frac{H(\Omega)}{R}\right) +$$

$$+ G\varrho_0 \int_{\Omega_{\psi_0}} \left[\widetilde{L^{-1}}(R, \psi, r')\Big|_{r'=R+H(\Omega)}^{R+H(\Omega')} - R^2 \frac{\tau(\Omega') - \tau(\Omega)}{\ell_0}\right] d\Omega', \qquad (4.2)$$

where Ω_{ψ_0} is a spherical cap of a small radius ψ_0 ($\leq 3°$) surrounding the computation point.

4.3 Planar approximation of distances

The planar approximation of distances (not to be confused with a planar approximation of the geoid) is based on the fact that the ratio H/R never exceeds the value of 1.4×10^{-3}. The planar approximation of distances is acceptable because it again produces an error of the same order of magnitude as the error in the spherical approximation of the geoid. Employing this approximation, quantities of the order of H/R are neglected with respect to 1. For example, the planar approximation of the spatial distance $L(R, \psi, H(\Omega'))$ between points (R, Ω) and $(R + H(\Omega'), \Omega')$ is simple to derive using eqn.(3.56):

$$L(R, \psi, R + H(\Omega')) \approx \sqrt{\ell_0^2 + H^2(\Omega')} . \qquad (4.3)$$

Approximating the function $\tau(\Omega)$ by $H(\Omega)$, $\tau(\Omega')$ by $H(\Omega')$, the Bouguer term $-2\pi G\varrho_0 H^2(\Omega)\left(1 + \frac{2}{3}\frac{H(\Omega)}{R}\right)$ becomes $-2\pi G\varrho_0 H^2(\Omega)$. Then the formula (4.2) reads

$$\delta V(R, \Omega) \doteq -2\pi G\varrho_0 H^2(\Omega) +$$

$$+ G\varrho_0 \int_{\Omega_{\psi_0}} \left[\widetilde{L^{-1}}(R, \psi, r')\Big|_{r'=R+H(\Omega)}^{R+H(\Omega')} - R^2 \frac{H(\Omega') - H(\Omega)}{\ell_0}\right] d\Omega' . \qquad (4.4)$$

Let us find now the planar approximation of the function $\widetilde{L^{-1}}(R, \psi, r')\Big|_{r'=R+H(\Omega)}^{R+H(\Omega')}$. Using eqn.(3.55), we have

$$\widetilde{L^{-1}}(R, \psi, r')\Big|_{r'=R+H(\Omega)}^{R+H(\Omega')} = \frac{R}{2}(1 + \frac{H(\Omega')}{R} + 3\cos\psi)L(R, \psi, R + H(\Omega')) -$$

$$-\frac{R}{2}\left(1 + \frac{H(\Omega)}{R} + 3\cos\psi\right)L(R,\psi,R+H(\Omega))+ \tag{4.5}$$

$$+\frac{R^2}{2}(3\cos^2\psi - 1)\ln\frac{R(1-\cos\psi)+H(\Omega')+L(R,\psi,R+H(\Omega'))}{R(1-\cos\psi)+H(\Omega)+L(R,\psi,R+H(\Omega))} \;.$$

Substituting for distances $L(R,\psi,R+H(\Omega'))$ and $L(R,\psi,R+H(\Omega))$ from eqn.(4.3), neglecting $H(\Omega)/R$ and $H(\Omega')/R$ with respect to 1 in the first two terms, and expressing the function $1-\cos\psi$ by means of the distance ℓ_0, we get

$$\widetilde{L^{-1}}(R,\psi,r')\Big|_{r'=R+H(\Omega)}^{R+H(\Omega')} \doteq \frac{R}{2}(1+3\cos\psi)\left(\sqrt{\ell_0^2+H^2(\Omega')}-\sqrt{\ell_0^2+H^2(\Omega)}\right)+$$

$$+\frac{R^2}{2}(3\cos^2\psi - 1)\ln\frac{\frac{\ell_0^2}{2R}+H(\Omega')+\sqrt{\ell_0^2+H^2(\Omega')}}{\frac{\ell_0^2}{2R}+H(\Omega)+\sqrt{\ell_0^2+H^2(\Omega)}} \;. \tag{4.6}$$

From eqn.(3.57), the function $1+3\cos\psi$ in terms of the distance ℓ_0 reads

$$1+3\cos\psi = 4 - \frac{\ell_0^2}{2R^2} \;. \tag{4.7}$$

Because of restricted integration over Ω_{ψ_0} in eqn.(4.4), we may neglect the term $\ell_0^2/2R^2$ with respect to 4 getting

$$1+3\cos\psi = 4 - O\left(1\times 10^{-4}\right) \;. \tag{4.8}$$

Within the same accuracy, we may further write

$$3\cos^2\psi - 1 \doteq 2 \;. \tag{4.9}$$

Both the last approximations are acceptable because they produce an error of the same order as the relative error of 3×10^{-3} of the spherical approximation used for topographical effects (sect.3.1). Using approximations (4.8) and (4.9), we get the planar approximation of the function $\widetilde{L^{-1}}(R,\psi,r')\Big|_{r'=R+H(\Omega)}^{R+H(\Omega')}$ in the form

$$\widetilde{L^{-1}}(R,\psi,r')\Big|_{r'=R+H(\Omega)}^{R+H(\Omega')} \doteq 2R\left(\sqrt{\ell_0^2+H^2(\Omega')}-\sqrt{\ell_0^2+H^2(\Omega)}\right)+$$

$$+R^2\ln\frac{\frac{\ell_0^2}{2R}+H(\Omega')+\sqrt{\ell_0^2+H^2(\Omega')}}{\frac{\ell_0^2}{2R}+H(\Omega)+\sqrt{\ell_0^2+H^2(\Omega)}} \;. \tag{4.10}$$

Finally, substituting the last formula into eqn.(4.4), the indirect topographical effect on potential may be approximated as

$$\delta V(R,\Omega) \doteq -2\pi G\varrho_0 H^2(\Omega) + GR^2\varrho_0 \int_{\Omega_{\psi_0}}\left[2\frac{\sqrt{\ell_0^2+H^2(\Omega')}-\sqrt{\ell_0^2+H^2(\Omega)}}{R}+\right.$$

$$\left.+\ln\frac{\frac{\ell_0^2}{2R}+H(\Omega')+\sqrt{\ell_0^2+H^2(\Omega')}}{\frac{\ell_0^2}{2R}+H(\Omega)+\sqrt{\ell_0^2+H^2(\Omega)}} - \frac{H(\Omega')-H(\Omega)}{\ell_0}\right]d\Omega' \;. \tag{4.11}$$

By an analogous way, we can show that the planar approximation of the integration kernels (3.53) and (3.54) of the direct topographical effect on gravity $\delta A(\Omega)$ read

$$\left.\frac{\partial L^{-1}(r,\psi,R)}{\partial r}\right|_{r=R+H(\Omega)} = -\frac{H(\Omega)}{\sqrt{[\ell_0^2 + H^2(\Omega)]^3}}, \qquad (4.12)$$

and

$$\left.\left.\frac{\partial \widetilde{L^{-1}}(r,\psi,r')}{\partial r}\right|^{R+H(\Omega')}_{r'=R+H(\Omega)}\right|_{r=R+H(\Omega)} =$$

$$= -R^2\left(\frac{1}{\sqrt{\ell_0^2 + [H(\Omega')-H(\Omega)]^2}} - \frac{1}{\ell_0}\right) + 2R\ln\frac{\frac{\ell_0^2}{2R} + H(\Omega') + \sqrt{\ell_0^2 + H^2(\Omega')}}{\frac{\ell_0^2}{2R} + H(\Omega) + \sqrt{\ell_0^2 + H^2(\Omega)}}. \qquad (4.13)$$

The direct topographical effect on gravity $\delta A(\Omega)$, c.f., eqn.(3.45), can then be simplified as

$$\delta A(\Omega) = -GR^2\varrho_0 \int_{\Omega_{\psi_0}} \left[\frac{1}{\sqrt{\ell_0^2 + [H(\Omega')-H(\Omega)]^2}} - \frac{1}{\ell_0} - \frac{H(\Omega)\,[H(\Omega')-H(\Omega)]}{\sqrt{[\ell_0^2 + H^2(\Omega)]^3}}\right.$$

$$\left. -\frac{2}{R}\ln\frac{\frac{\ell_0^2}{2R} + H(\Omega') + \sqrt{\ell_0^2 + H^2(\Omega')}}{\frac{\ell_0^2}{2R} + H(\Omega) + \sqrt{\ell_0^2 + H^2(\Omega)}}\right] d\Omega'. \qquad (4.14)$$

4.4 The difference between spherical and planar approximation of topographical effects

Using the planar approximation of the geoid, Heck (1993) derived that the change of the gravitational potential due to the condensation (for $H \geq 0$) is equal to (ibid., eqn.(9))

$$\delta V_p(R,\Omega) = -\pi G\varrho_0 H^2(\Omega) +$$

$$+GR^2\varrho_0 \int_{\Omega_{\psi_0}} \left[\ln\frac{H(\Omega') + \sqrt{\ell_0^2 + H^2(\Omega')}}{H(\Omega) + \sqrt{\ell_0^2 + H^2(\Omega)}} - \frac{H(\Omega') - H(\Omega)}{\ell_0}\right] d\Omega'. \qquad (4.15)$$

He also showed (ibid., eqn.(10)) that the change of the gravitational attraction due to condensation is

$$\delta A_p(\Omega) = \qquad (4.16)$$

$$= -GR^2\varrho_0 \int_{\Omega_{\psi_0}} \left[\frac{1}{\sqrt{\ell_0^2 + [H(\Omega')-H(\Omega)]^2}} - \frac{1}{\ell_0} - \frac{H(\Omega)\,[H(\Omega')-H(\Omega)]}{\sqrt{[\ell_0^2 + H^2(\Omega)]^3}}\right] d\Omega'.$$

Comparing these equations with eqns.(4.11) and (4.14), we can see that $\delta V_p(R,\Omega)$ differs from $\delta V(R,\Omega)$ and $\delta A_p(\Omega)$ differs from $\delta A(\Omega)$. Let us discuss the differences. The Bouguer term $-2\pi G\varrho_0 H^2(\Omega)$ in eqn.(4.11) derived from a spherical approximation of the geoid has a magnitude twice as large as the equivalent term in Heck's formula (4.15). The difference, $-\pi G\varrho_0 H^2(\Omega)$, is always negative and may reach up to -5 metres in high mountains. Therefore, eqn. (4.15) derived from the planar approximation of the geoid is biased.

Equation (4.15) also differs from eqn.(4.11) in the terrain roughness term. We cannot neglect the term $\ell_0^2/2R$ with respect to $H(\Omega)$ in the logarithmic function because for distances over $1°$ the magnitude of the term $\ell_0^2/2R$ may be comparable with $H(\Omega)$ or $H(\Omega')$. Moreover, the logarithmic function is completely neglected in planar approximation $\delta A_p(\Omega)$. All these differences may become crucial for a rugged mountainous terrain because the terrain roughness term affects the geoid in the range of one metre.

Comparison of $\delta A(\Omega)$ with the expression used by, e.g., Wang and Rapp (1990), or Sideris and Forsberg (1990),

$$\widehat{\delta A}(\Omega) = \frac{1}{2}GR^2\rho_0 \int_{\Omega_0} \frac{[H(\Omega') - H(\Omega)]^2}{\ell_0^3}d\Omega' , \qquad (4.17)$$

reveals even more serious problems. The derivation of this formula necessitates that the Pellinen (1962) credo of linear dependence of free-air anomalies on heights be invoked. This makes the validity of eqn.(4.17) for computing the direct topographical effect in the context of geoid computations questionable (Martinec et al., 1993).

4.5   Conclusion

All existing theories of topographical effects in Helmert's 2nd condensation technique are based on the concept of planar approximation of the geoid. The geoid is considered as an infinite plane and topographical masses are condensed onto this plane. This approximation describes the actual situation only very roughly.

The theoretical study in this section was motivated by the above inconsistency in the description of the problem. Modelling the geoid by a sphere removes infinite potentials appearing in planar approximation and describes the actual situation much more accurately. In order to compare our results with those recently derived by Heck (1993), we have started with the spherical formulae for both the direct and indirect topographical effects and have employed additional simplifications. We have shown that the planar formula for the indirect effect is biased in a term corresponding to the indirect topographical effect of the spherical Bouguer shell and that this bias may reach -5 m in high mountains. We have also demonstrated that the planar approximation of terrain roughness terms is of questionable accuracy for a mountainous region. We can conclude that the expressions based on the planar model of the geoid are uniformly 'blind' to all spherical effects and they cannot be used for evaluation of the topographical effects if one-decimetre accuracy in geoid determination is desired.

Chapter 5

Taylor series expansion of the Newton kernel

The strong gravitational field induced by topographical masses (Martinec, 1991; Martinec, 1994b) poses a question of how the gravitational effect of compensated masses should be computed numerically. One possible way is as follows (Moritz, 1966, 1968; Wichiencharoen, 1982; Vaníček and Kleusberg, 1987; Wang and Rapp, 1990; Sideris and Forsberg, 1990; Heck, 1993; Martinec and Vaníček, 1994a,b): first, the Newton integral is formulated for the potential of topographical masses. Then its kernel is expanded by means of Taylor's series with respect to the radial coordinate at a point on the geoid. Then integration over the vertical coordinate is carried out analytically. Finally, the singularity of each individual term of Taylor series is removed. The zero-degree term of the Taylor expansion corresponds to the potential of the condensation layer. It describes the behaviour of the bulk of the potential of the topographical masses and is usually included into an unknown anomalous gravitational potential. The potential generated by higher-order terms of the Taylor series can therefore be viewed as corrections to the condensation layer potential; provided that the density of the topographical masses is known, the Newton integral can be used to compute this residual potential.

A questionable point of the above procedure, pointed out, e.g., by Heck (1993), is whether the Taylor series converges or not, and if so, how many terms of the series should be taken into consideration to describe the gravitational potential with a prescribed accuracy. It is the usual practice in geodesy (Moritz, 1968; Vaníček and Kleusberg, 1987; Sideris and Forsberg, 1990; Forsberg and Sideris, 1993) to take only a few first terms of the Taylor series (most often only the first three) and assume that the rest of the series may be neglected. This seems to be a good enough approximation for a flat terrain when the topographical heights can be taken on a grid of a large step size (e.g. 0.5 degree). Then the dummy point in the numerical integration (of the Newton integral) never comes too close to the computation point and the magnitudes of higher order terms of the Taylor series remain small.

The problems appear when the gravitational potential of the topographical masses is computed in a rugged terrain. For such a case, a grid of topographical heights has to be fairly dense to express the ruggedness of the terrain. The dummy point in the Newton integration than comes close to the computation point and the magnitudes of higher order Taylor terms may increase faster than the magnitudes of lower order terms. As a result, higher order terms become dominant and the series no longer converges (Martinec et al., 1996).

In this chapter we are looking for the condition under which the Taylor series expansion converges. Having found this condition, we will treat a question of how many terms of the Taylor series should be taken into consideration to describe the gravitational effect of topographical masses with a prescribed accuracy. The correct analytical formula (3.52) for evaluation of the radial integral of the Newton kernel yields an independent tool for investigating the convergence of the Taylor series discussed above. We will see that this series has the most unfavorable behaviour when it is evaluated at a point on the geoid. Therefore, the numerical tests will be carried out for points on the geoid. This, incidently, corresponds to the evaluation of the primary indirect topographical effect on the geoid. This technique works, of course, the same way for any other point and can thus be used in the evaluation of the direct topographical effect on gravity as well elsewhere.

To begin with, let us expand the kernel $r'^2/L(r, \psi, r')$ of the Newton volume integral (3.13) into a Taylor series. As we have shown in sect.3.2, the singular point $\psi = 0$ can be left out from the integration domain of integral (3.13). Consequently, the distance $L(r, \psi, r')$ is never equal to zero, and the reciprocal distance $1/L(r, \psi, r')$ is always bounded. The function $r'^2/L(r, \psi, r')$ (of variable r') may be therefore expanded into a Taylor series as

$$\frac{r'^2}{L(r, \psi, r')} = R^2 \sum_{i=0}^{\infty} \frac{1}{i!} M_i(r, \psi, R) \left(\frac{r' - R}{R}\right)^i, \tag{5.1}$$

where new integration kernels $M_i(r, \psi)$, $i \geq 0$, are again isotropic and are defined as

$$M_i(r, \psi, R) := R^{i-2} \frac{\partial^i}{\partial r'^i} \left(\frac{r'^2}{L(r, \psi, r')}\right)\Big|_{r'=R}. \tag{5.2}$$

5.1 The problem of the convergence of Taylor series expansion

The problem of using the Taylor series expansion (5.1) in topographical effect computation is demonstrated most transparently in the case of the primary indirect topographical effect on potential. In this case the topographical potential V^t, and therefore the integration kernels M_i are to be evaluated on the geoid, i.e., at points (R, Ω). As shown in Appendix 5.4, all the kernels $M_i(R, \psi, R)$ are singular at the point $\psi = 0$. Fortunately, we have learned by analysing eqn.(3.13) that there is no need to evaluate the integration kernel $M_i(R, \psi, R)$ at the point

$\psi = 0$. Nevertheless, the kernels have to be evaluated in the immediate neighbourhood of the computation point $\psi = 0$, and therefore, we have to ask about the convergence radius of the series (5.1). Namely, we intend to exchange the order of integration over r' and Taylor summation over i in the Newton integral (3.13).

To answer this question, we employ eqn.(A.5.14); for $r = R$ it yields

$$M_i(R, \psi, R) = \frac{1}{\ell_0} \sum_{s=1}^{i-1} \frac{i!(i-2)!}{(i-s-1)!(s-1)!} \left(\frac{R}{\ell_0}\right)^{i+1-s} \times$$

$$\times \sum_{t=0}^{i+1-s} (-1)^{\frac{1}{2}(3i+1-s+t)} \frac{(i+2-s-t)!!(i-s+t)!!}{(i+2-s-t)!\,t!} \left(\frac{\ell_0}{2R}\right)^t , \tag{5.3}$$

where ℓ_0 stands for the distance between points (R, Ω) and (R, Ω'), c.f., eqn.(3.57). The type of the singularity of the kernel $M_i(R, \psi, R)$ at the point $\psi = 0$ is given by the highest power of the reciprocal distance $1/\ell_0$. Inspecting eqn.(5.3), we can see that the highest power of $1/\ell_0$ occurs when summation indices s and t are small, i.e., $s = 1$, and $t = 0$ or $t = 1$ (subject to the constraint that $i - s + t + 1$ be an even number). Considering $s = 1$ in this constraint, we can see that t must be of the same parity as i, i.e., $t = 0$ when i is an even number, and $t = 1$ when i is an odd number.

To begin with, let us consider the case when i is an *even* number ($i = 0, 2, ...$). From eqn.(5.3) we can see that in the vicinity of the point $\psi = 0$, (i.e., when $\psi \leq \varepsilon$, and ε is a small positive number), the kernel $M_i(R, \psi, R)$ behaves as

$$M_i(R, \psi \leq \varepsilon, R) \approx (-1)^{\frac{3}{2}i} [(i-1)!!]^2 \frac{R^i}{\ell_0^{i+1}} . \tag{5.4}$$

This formula shows that the higher the degree of the kernel $M_i(r, \psi)$, the stronger its singularity at the point $\psi = 0$.

Equation (5.1) represents the integration kernel $r'^2/L(r, \psi, r')$ as a power series of variable r'. The convergence radius d of this series in the vicinity of the point $\psi = 0$ is

$$d = \frac{1}{D} \tag{5.5}$$

where (i is still an even number)

$$D^2 := \lim_{i \to \infty} \left| \frac{\dfrac{M_{i+2}(R, \psi \leq \varepsilon, R)}{(i+2)!\,R^{i+2}}}{\dfrac{M_i(R, \psi \leq \varepsilon, R)}{i!\,R^i}} \right| . \tag{5.6}$$

Substituting from eqn.(5.4), we get

$$D^2 = \lim_{i \to \infty} \frac{(i+1)^2}{(i+2)(i+3)} \left(\frac{1}{\ell_0}\right)^2 . \tag{5.7}$$

Since

$$\lim_{i\to\infty} \frac{(i+1)^2}{(i+2)(i+3)} = 1 , \tag{5.8}$$

$D = 1/\ell_0$, and the convergence radius d of the series (5.1) is ℓ_0. In other words, in the vicinity of the point $\psi = 0$ the series (5.1) converges provided that r' is taken from the interval

$$\boxed{r' \in (R - \ell_0, R + \ell_0) .} \tag{5.9}$$

The same convergence radius may be found for odd degrees i.

In the case when the kernels $M_i(R, \psi, R)$ are evaluated for such ψ's that $\ell_0 > H(\Omega')$, the Taylor series (5.1) converges within the whole integration domain $r' \in (R, R + H(\Omega'))$, and the order of the integration over r' and the Taylor summation over i in the Newton integral (3.13) can be exchanged. On the other hand, computing the kernels $M_i(R, \psi, R)$ at a smaller distance ψ, such that $\ell_0 < H(\Omega')$, the Taylor series (5.1) diverges, and we cannot exchange the order of the integration over r' and the Taylor summation over i.

As a consequence, special care has to be paid to numerical computation of the Newton integral for V^t when the Taylor series is employed. The topographical heights are usually given on a regular grid $(R\Delta\phi, R\cos\phi\Delta\lambda)$, where $\Delta\phi$ and $\Delta\lambda$ are steps in latitude and longitude, respectively. If both the steps $R\Delta\phi$ and $R\cos\phi\Delta\lambda$ are larger than the largest topographical height in the area of interest, then the values of the distance ℓ_0 over the topographical grid are always larger than $H(\Omega')$, and the Taylor series (5.1) converges at all the points on the topographical grid. Such a situation usually occurs in a flat terrain.

The problem arises when the topographical potential $V^t(r, \Omega)$ is to be evaluated at mountainous terrain. In this case, the sampling of the topographical heights has to be fairly dense (e.g., with a step of 1 km) to express the ruggedness of the terrain; the dummy point in the Newton integral then may come very close to the computation point so that $\ell_0 < H(\Omega')$. At this point the higher order terms of the Taylor series become larger than the low degree terms and the series (5.1) diverges. Moreover, as eqn.(5.4) shows, individual terms of Taylor series change the sign. To use such an oscillating and divergent series for the evaluation of the Newton integral becomes impossible; we will see later that it cannot be used for an accurate geoid determination.

5.2 The Taylor expansion of the terrain roughness term

Now, let us substitute the Taylor series (5.1) into the radially integrated Newton kernel $\widetilde{L^{-1}}(r, \psi, r')$, c.f., eqn.(3.35), and assume that the integration domain over r' lies within the convergence interval (5.9). Then, we can exchange the order of integration and summation, and integrate each individual term of the Taylor series separately. The results can be written in the following form

$$\widetilde{L^{-1}}(r,\psi,r')\Big|_{r'=R}^{R+H(\Omega')} = \sum_{i=0}^{\infty} K_i(r,\psi,H(\Omega'))\,, \qquad (5.10)$$

where the Taylor series terms K_i read

$$K_i(r,\psi,H(\Omega')) := \frac{R^3}{(i+1)!}\left(\frac{H(\Omega')}{R}\right)^{i+1} M_i(r,\psi,R)\,. \qquad (5.11)$$

Now, we are ready to evaluate the terrain roughness term $\delta V^{\sigma,R}(R,\Omega)$ for the primary indirect topographical effect on potential. By formulae (3.51) and (3.30), $\delta V^{\sigma,R}(R,\Omega)$ reads

$$\delta V^{\sigma,R}(R,\Omega) = G\int_{\Omega_0}\left[\overline{\varrho}(\Omega')\,\widetilde{L^{-1}}(R,\psi,r')\Big|_{r'=R}^{R+H(\Omega')} - \overline{\varrho}(\Omega)\,\widetilde{L^{-1}}(R,\psi,r')\Big|_{r'=R}^{R+H(\Omega)} - \right.$$

$$\left. -R^2\,\frac{\overline{\varrho}(\Omega')H(\Omega') - \overline{\varrho}(\Omega)H(\Omega)}{L(R,\psi,R)}\right]d\Omega'\,. \qquad (5.12)$$

Substituting the Taylor series (5.10), we get

$$\delta V^{\sigma,R}(R,\Omega) = G\int_{\Omega_0}\left[\sum_{i=0}^{\infty}[\overline{\varrho}(\Omega')K_i(R,\psi,H(\Omega')) - \overline{\varrho}(\Omega)K_i(R,\psi,H(\Omega))] - \right.$$

$$\left. -R^2\,[\overline{\varrho}(\Omega')H(\Omega') - \overline{\varrho}(\Omega)H(\Omega)]M_0(R,\psi,R)\right]d\Omega'\,. \qquad (5.13)$$

By eqn.(5.11), the kernel K_0 can be expressed by means of the kernel M_0, and the expression (5.13) can further be simplified as

$$\delta V^{\sigma,R}(R,\Omega) = G\int_{\Omega_0}\sum_{i=1}^{\infty}[\overline{\varrho}(\Omega')K_i(R,\psi,H(\Omega')) - \overline{\varrho}(\Omega)K_i(R,\psi,H(\Omega))]\,d\Omega'\,. $$
$$(5.14)$$

In practice, the infinite series in the last equation is truncated at a degree i_{max} and the correct value of $\delta V^R(R,\Omega)$ is approximated by an estimate $\widehat{\delta V^R}(R,\Omega)$:

$$\widehat{\delta V^{\sigma,R}}(R,\Omega) = G\int_{\Omega_0}\sum_{i=1}^{i_{max}}[\overline{\varrho}(\Omega')K_i(R,\psi,H(\Omega')) - \overline{\varrho}(\Omega)K_i(R,\psi,H(\Omega))]\,d\Omega'\,. $$
$$(5.15)$$

For example, Wichiencharoen (1982), Vaníček and Kleusberg (1987), and Wang and Rapp (1990) consider only the terms with $i = 2$.

5.3   Numerical computations

5.3.1   The Taylor kernels K_i

The correct formula (5.12) for $\delta V^{\sigma,R}(R,\Omega)$ with an analytical expression (3.52) for the integration kernel $\widetilde{L^{-1}}(r,\psi,r')$ provides us with a tool for testing the accuracy

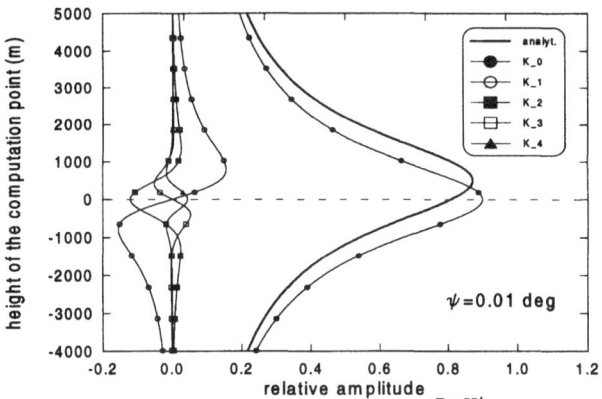

Figure 5.1: Integration kernels $\widetilde{L^{-1}}(R + H, \psi, r')\big|_{r'=R}^{R+H'}$ computed by analytical formula (3.52) (curve denoted as 'analyt.') and the Taylor series expansion kernels $K_i(R + H, \psi, H')$ for $i = 0, ..., 4$. The height H of the computation point ranges in the interval $(-4000, 5000)$ m, the height H' of an integration point is 1000 m. The angular distance ψ between the computation point and the integration point is $0.01°$.

of the estimate $\delta \widehat{V^{\sigma,R}}(R, \Omega)$. To demonstrate the problem with the Taylor series convergence, we computed the definite integral $\widetilde{L^{-1}}(R + H(\Omega), \psi, r')\big|_{r=R}^{R+H(\Omega')}$ using the analytical expression (3.52) and compared its behaviour with that of the first 5 terms of the Taylor series (5.10). In Figure 5.1, we have taken the height $H(\Omega')$ of the dummy point of integration to be 1 km, the angular distance ψ to be $0.01°$ (which corresponds to the horizontal distance ℓ_0 of about 1.1 km), and the height $H(\Omega)$ of the computation point has been varied from -1 km to 5 km ($H(\Omega) = 0$ means that the computation point lies on the geoid).

From Figure 5.1 we can observe that:

- The magnitude of the Taylor series kernels K_i decreases with increasing order i; in other words the Taylor series (5.10) converges which confirms our theoretical conclusion because $\ell_0 > H(\Omega')$.

- The kernels K_i change their signs for the computation point near the geoid, which means that the Taylor series (5.10) oscillates for points near the geoid.

- For a computation point on the geoid, the magnitudes of odd-degree kernels K_i are very small (but not equal to zero) compared with those of even-degree kernels K_i. This means that the primary indirect topographical effect on potential can be reasonably modelled by even degree kernels K_i only.

Figure 5.2 shows a similar situation as Figure 5.1, where the only change is that the angular distance ψ between the computation and dummy points is one half of the above, i.e., the horizontal distance ℓ_0 is 0.55 km. As already shown

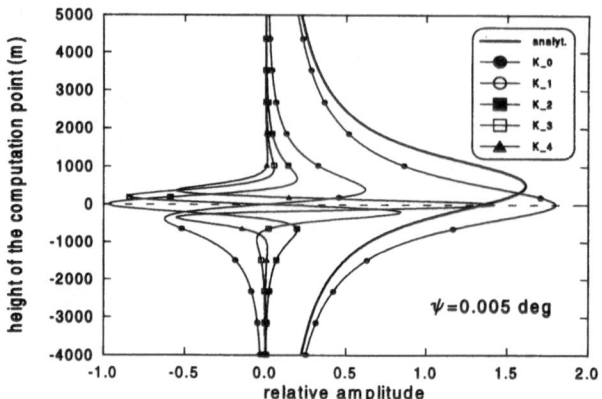

Figure 5.2: The same as Figure 5.1 but for $\psi = 0.005°$.

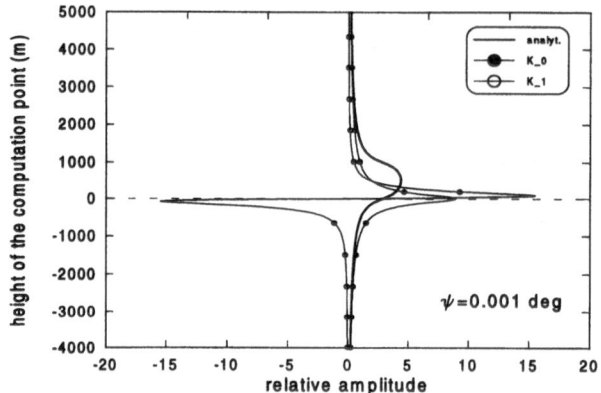

Figure 5.3: The same as Figure 5.1 but for $\psi = 0.001°$.

theoretically, the kernels K_i grow with increasing order i at the point on the geoid because $\ell_0 < H(\Omega')$, and, consequently, the Taylor series (5.10) does not converge. Figure 5.2 demonstrates this fact very clearly.

An even more drastic example of the divergence of the series (5.10) is plotted in Figure 5.3. Here the angular distance ψ is equal to $0.001°$ which corresponds to the horizontal distance $\ell_0 = 0.11$ km. We have plotted only the first two kernels K_0 and K_1 because the magnitude of the higher order kernels grows too fast.

5.3.2 The primary indirect topographical effect on potential

Particular contributions of the Taylor series kernels K_i, to the terrain roughness term $\delta \widehat{V^{\sigma,R}}(R,\Omega)$ were studied over an area of $1.65° \times 1.25°$, $(51.20° \leq \phi \leq 52.85°; 242.7 \leq \lambda \leq 243.95°)$ in Western Canada. The area covers a particularly rugged part of the Rocky Mountains, the chain of Columbia Mountains. The

Table 5.1: Corrections to the geoidal heights (in metres) due to the terrain roughness term $\delta\widehat{V^{\sigma,R}}(R,\Omega)$ expressed by Taylor series kernels K_i, $i=1,...4$.

term	min	max
K_1	-0.006	0.003
K_2	-0.222	0.394
K_3	0.000	0.000
K_4	-1.14	0.906

topographical heights range from 0 to 3573 m, and were sampled as means of $30'' \times 60''$ cells.

Table 5.1 shows the extreme values of corrections to geoidal heights due to the terrain roughness term $\delta\widehat{V^{\sigma,R}}(R,\Omega)$, c.f., eqn.(5.15), induced by kernels K_i, $i=1,...,4$. The correct values, computed by means of the analytical formula (3.52), are shown in Figure 5.4; they fall within the interval (-0.086, 0.197) m.

For the sake of completeness, we remark that the Bouguer term standing on the right-hand side of eqn.(3.51) for the same region gives values ranging from -1.46m to 0m for the $30'' \times 60''$ grided heights and from -0.71m to 0m for the $5' \times 5'$ grided heights. This demonstrates how sensitive the results are to the selected grid step size.

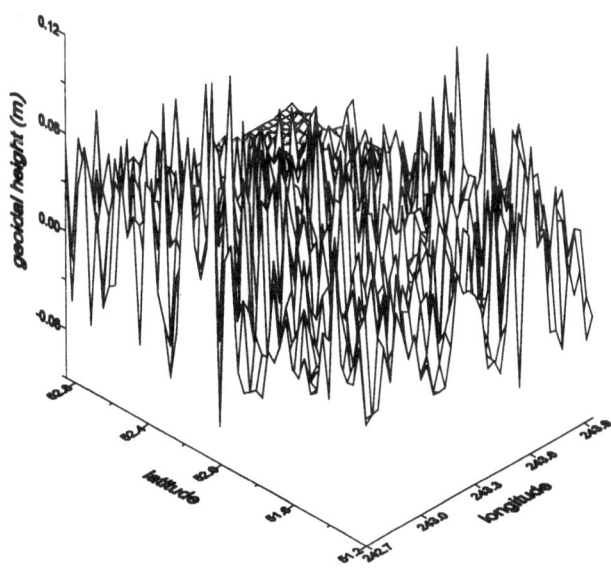

Figure 5.4: Geoidal heights (in metres) induced by the terrain roughness term $\delta V^{\sigma,R}(R,\Omega)$ of the primary indirect topographical effect on the geoid over the area of Columbia Mountains.

Inspecting Table 5.1 and Figure 5.4, we can draw the following conclusions:

- The magnitude of $\widehat{\delta V^{\sigma,R}}(R,\Omega)$ induced by kernels K_i with odd orders are small enough to be neglected for a 'decimetre geoid' computation in Canada.

- The magnitudes of $\widehat{\delta V^{\sigma,R}}(R,\Omega)$ induced by kernels K_i with even orders increase with increasing order i; the Taylor series expansion (5.10) diverges at points with extreme heights.

- Wang and Rapp (1990) modelled the terrain roughness term $\delta V^{\sigma,R}(R,\Omega)$ by the contribution generated by the kernel K_2 using the topographical grid size 30". Comparing these values with the correct ones, we can see that the K_2 term overestimates the terrain corrections $\delta V^{\sigma,R}(R,\Omega)$ about two times. This means that errors in geoidal heights due to the K_2 term approximation are about two decimetres over the area under study. These errors cannot be reduced by adding higher order terms of the Taylor series, e.g., by adding the K_4 contribution, because of the divergence of this series.

- We have also computed the terrain roughness term $\delta V^{\sigma,R}(R,\Omega)$ from heights averaged on a sparser topographical grid $5' \times 5'$ (not shown here). The resulting values fall within the interval (-0.03, 0.03) m. Comparing these values with those on Figure 5.4, we can see that the differences are more than one decimetre. This is caused by the fact that the mean $5' \times 5'$ digital terrain model is smoother than the $30'' \times 60''$ model. This fact tells us that a grid size of $5'$ in a mountainous terrain such as the Rocky Mountains is not sufficiently dense to express the irregularities of the terrain and thus does not reveal properly the contribution to geoidal height due to terrain height variations.

5.4 Conclusion

In this chapter we have concentrated on the problem of convergence of the Taylor series of the Newton gravitational kernel on a dense topographical grid used for modelling a rugged terrain. We have shown that for the primary indirect topographical effect on potential, this expansion converges if the horizontal distance between the computation point and the dummy integration point is greater than the topographical elevation. In fact, this constrains the sampling of topographical heights. On the one hand, there is a need to discretize the heights of a rugged terrain as densely as possible to model the terrain irregularities as accurately as possible and, on the other hand, the discretization step cannot be smaller than the largest topographical elevation in order for the Taylor series to converge. On a numerical example we have demonstrated that the Rocky Mountains is one of the areas where these contradicting requirements on the digitization step are not possible to be satisfied simultaneously.

If the gravitational potential of topographical masses at the point on the geoid is approximated by the zeroth and quadratic terms of the Taylor series, as it is usually done, then such an approximation is biased and the bias may easily reach decimetres. We have shown that for the Columbia Mountains these errors reach in fact two decimetres.

Our conclusion is that for a rugged terrain the approximate formulae based on the Taylor series expansion have to replaced by the correct (analytical) formula (3.52) which removes the problem of divergence of the Taylor series. In sect.3.8.1 we have shown the very stable behaviour of this formula in the vicinity of the computation point ($\psi = 0$). However, considering this correct formula, the application of the fast Fourier technique for computing the gravitational potential of the topographical masses is not as straightforward as in the case when the Taylor series expansion is employed (see also Heck, 1993). Further investigation of this problem has to be carried out.

A.5 Integration kernels $M_i(r, \psi, R)$

The integration kernels $M_i(r, \psi, R)$ play a fundamental role in the Taylor series expansion (5.1). Let us therefore derive some useful properties of theirs.

A.5.1 Spectral form

Let us start by expressing the integration kernel $M_i(r, \psi, R)$ as a series of Legendre polynomials $P_j(\cos \psi)$. To begin with, let us consider that $r > r'(= R)$; the Newton kernel may then be expanded into a uniformly convergent series as

$$\frac{1}{L(r, \psi, r')} = \frac{1}{r} \sum_{j=0}^{\infty} \left(\frac{r'}{r}\right)^j P_j(\cos \psi) \,. \tag{A.5.1}$$

Multiplying eqn.(A.5.1) by r'^2 and differentiating the result i-times with respect to r', we get

$$\frac{\partial^i}{\partial r'^i} \left(\frac{r'^2}{L(r, \psi, r')}\right) = (r')^{-i+1} \sum_{j=0}^{\infty} (j+2)(j+1)...(j+2-i+1) \left(\frac{r'}{r}\right)^{j+1} P_j(\cos \psi) \,. \tag{A.5.2}$$

It is possible to change the order of summation and differentiation because the series on the right-hand side is uniformly convergent. Inserting eqn.(A.5.2) into (5.2), we obtain for all $i \geq 0$:

$$M_i(r, \psi, R) = \frac{1}{R} \sum_{j=0}^{\infty} (j+2)(j+1)...(j+2-i+1) \left(\frac{R}{r}\right)^{j+1} P_j(\cos \psi) =$$

$$= \frac{i!}{R} \sum_{j=0}^{\infty} \binom{j+2}{i} \left(\frac{R}{r}\right)^{j+1} P_j(\cos \psi) \,. \tag{A.5.3}$$

By an analogous manner, we can show that for $r \leq R$, it holds

$$M_i(r, \psi, R) = \frac{i!}{R} \sum_{j=0}^{\infty} \binom{-j+1}{i} \left(\frac{r}{R}\right)^j P_j(\cos \psi) \,. \tag{A.5.4}$$

A.5.2   Recurrence formula

The integration kernels $M_i(r, \psi, R)$ may be obtained by means of a recurrence formula. Multiplying eqn.(A.5.3) by r^{i-1} and differentiating with respect to r, we get

$$\frac{\partial}{\partial r}\left(r^{i-1} M_i(r, \psi, R)\right) = -\frac{i!}{R}r^{i-2} \sum_{j=0}^{\infty}\left(\begin{array}{c} j+2 \\ i \end{array}\right)(j+2-i)\left(\frac{R}{r}\right)^{j+1} P_j(\cos\psi) .$$

(A.5.5)

But, it can be shown that:

$$i!\left(\begin{array}{c} j+2 \\ i \end{array}\right)(j+2-i) = (i+1)!\left(\begin{array}{c} j+2 \\ i+1 \end{array}\right) ,$$

(A.5.6)

and eqn.(A.5.5) becomes

$$\frac{\partial}{\partial r}\left(r^{i-1} M_i(r, \psi, R)\right) = -r^{i-2}\frac{(i+1)!}{R} \sum_{j=0}^{\infty}\left(\begin{array}{c} j+2 \\ i+1 \end{array}\right)\left(\frac{R}{r}\right)^{j+1} P_j(\cos\psi) .$$ (A.5.7)

Taking into account eqn.(A.5.3) (for $i+1$), we get

$$\boxed{M_{i+1}(r, \psi, R) = -\frac{1}{r^{i-2}}\frac{\partial}{\partial r}\left(r^{i-1} M_i(r, \psi, R)\right) .}$$

(A.5.8)

The initial value for the recurrence process is $M_0(r, \psi, R)$. Putting $i = 0$ in eqn.(A.5.3), we have:

$$M_0(r, \psi, R) = \frac{1}{\ell} ,$$

(A.5.9)

where the symbol ℓ denotes the spatial distance between points (r, Ω) and (R, Ω'), i.e.,

$$\ell := L(r, \psi, R) = \sqrt{r^2 + R^2 - 2rR\cos\psi} .$$

(A.5.10)

By means of eqn.(A.5.4), it is not difficult to show that the recurrence formula for the kernels $M_i(r, \psi, R)$, when $r < R$, has the same form as eqn.(A.5.8). Hence, the recurrence formula (A.5.8) is valid for both the cases $r > R$ and $r < R$.

A.5.3   Spatial form

The integration kernels $M_i(r, \psi, R)$ may be further expressed as power series of the reciprocal distance $1/\ell$. This representation is helpful for understanding the behaviour of the kernel in the vicinity of the computation point ($\psi = 0$). Applying the relation (A.5.8) i-times recurrently, we get (for details see sect.A.5.6)

$$M_i(r, \psi, R) = (-1)^i \sum_{s=1}^{i-1} \frac{i!(i-2)!}{(i-s+1)!(i-s-1)!(s-1)!} r^{i+1-s}\frac{\partial^{i+1-s} M_0(r, \psi, R)}{\partial r^{i+1-s}} .$$

(A.5.11)

We note that this expression is valid only for $i \geq 2$. Taking into account (A.5.9), the series in eqn.(A.5.11) consists of higher-order radial derivatives of the reciprocal distance $1/\ell$. These may be expressed as

$$\frac{\partial^k}{\partial r^k}\left(\frac{1}{\ell}\right) = \sum_{t=0}^{k}(-1)^{\frac{k+t}{2}}\frac{(k-t+1)!!(k+t-1)!!}{(k-t+1)!}\frac{k!}{t!}\frac{(r-R\cos\psi)^t}{\ell^{k+t+1}}\,, \qquad (A.5.12)$$

where the summation must be taken over such t's for which $k+t$ is an even number. Inserting (A.5.12) into (A.5.11), and denoting by z the quantity

$$z := r - R\cos\psi\,, \qquad (A.5.13)$$

we get ($\forall i > 1$):

$$\boxed{\begin{aligned}
M_i(r,\psi,R) &= \frac{1}{\ell}\sum_{s=1}^{i-1}\frac{i!(i-2)!}{(i-s-1)!(s-1)!}\left(\frac{r}{\ell}\right)^{i+1-s}\times \\
&\times \sum_{t=0}^{i+1-s}(-1)^{\frac{1}{2}(3i+1-s+t)}\frac{(i+2-s-t)!!(i-s+t)!!}{(i+2-s-t)!\,t!}\left(\frac{z}{\ell}\right)^t\,,
\end{aligned}} \qquad (A.5.14)$$

and $i - s + t + 1$ must be an even number. The last equation represents the integration kernels $M_i(r,\psi,R)$ as finite power series of the reciprocal distance $1/\ell$.

The kernels $M_i(r,\psi,R)$ for the first few degrees are (the kernel $M_0(r,\psi,R)$ is given by eqn.(A.5.9)):

$$M_1(r,\psi,R) = \frac{1}{\ell} + \frac{rz}{\ell^3}\,, \qquad (A.5.15)$$

$$M_2(r,\psi,R) = -\frac{r^2}{\ell^3} + \frac{3r^2z^2}{\ell^5}\,, \qquad (A.5.16)$$

$$M_3(r,\psi,R) = \frac{3r^2}{\ell^3}\left(1 - \frac{3rz}{\ell^2} - \frac{3z^2}{\ell^2} + \frac{5rz^3}{\ell^4}\right)\,, \qquad (A.5.17)$$

$$M_4(r,\psi,R) = \frac{3r^2}{\ell^3}\left(-4 + \frac{3r^2}{\ell^2} + \frac{24rz}{\ell^2} + \frac{12z^2}{\ell^2} - \frac{40rz^3}{\ell^4} - \frac{30r^2z^2}{\ell^4} + \frac{35r^2z^4}{\ell^6}\right)\,. \qquad (A.5.18)$$

A.5.4 Singularity at the point $\psi = 0$

An important property of the integration kernel $M_i(r,\psi,R)$ is its behaviour in the vicinity of the point $\psi = 0$. At this point, the distance ℓ is equal to the height H of the computation point above the geoid,

$$\ell|_{\psi=0} = r - R = H\,. \qquad (A.5.19)$$

The integration kernel $M_i(r,\psi,R)$ may then be written as

$$\forall i > 1: \qquad M_i(R+H,0,R) = \frac{1}{H}\sum_{s=1}^{i-1}a_{is}\left(\frac{R+H}{H}\right)^{i+1-s}\,, \qquad (A.5.20)$$

and

$$M_1(R + H, 0, R) = \frac{1}{H} + \frac{R + H}{H^2} , \tag{A.5.21}$$

where the coefficients a_{is} are readily obtained from eqn.(A.5.14) as

$$a_{is} := \frac{i!(i-2)!}{(i-s-1)!(s-1)!} \sum_{t=0}^{i+1-s} (-1)^{\frac{1}{2}(3i+1-s+t)} \frac{(i+2-s-t)!!(i-s+t)!!}{(i+2-s-t)!\,t!} , \tag{A.5.22}$$

The summation is again taken over those t's for which $i - s + t + 1$ is an even number. Eqns.(A.5.20) and (A.5.21) show that if the height of the computation point goes to zero, the integral kernel $M_i(R+H, 0, R)$ goes to infinity as $1/H^{i+1}$.

A.5.5   Angular integrals

Now, let us evaluate the angular integrals

$$\int_{\Omega_0} M_i(r, \psi, R) d\Omega' . \tag{A.5.23}$$

We begin with $i = 1$. Applying Poisson's integral (Heiskanen and Moritz, 1967, eq.(1-88)) to a function R/r, we get

$$\frac{1}{4\pi} \int_{\Omega_0} \sum_{j=0}^{\infty} (2j+1) \left(\frac{R}{r}\right)^{j+1} P_j(\cos\psi) d\Omega' = \frac{R}{r} . \tag{A.5.24}$$

The angular integral (A.5.23) may then be expressed in terms of (A.5.24) as

$$\int_{\Omega_0} M_1(r, \psi, R) d\Omega' = \frac{1}{2R} \int_{\Omega_0} \sum_{j=0}^{\infty} (2j+1) \left(\frac{R}{r}\right)^{j+1} P_j(\cos\psi) d\Omega' +$$

$$+ \frac{3}{2R} \int_{\Omega_0} \sum_{j=0}^{\infty} \left(\frac{R}{r}\right)^{j+1} P_j(\cos\psi) d\Omega' . \tag{A.5.25}$$

Since the series over j in the last integral is uniformly convergent (we assume that $r > R$), we may change the order of integration and summation. Then using the orthogonality relation for the Legendre polynomials,

$$\frac{1}{4\pi} \int_{\Omega_0} P_j(\cos\psi) d\Omega' = \delta_{j0} , \tag{A.5.26}$$

we get

$$\int_{\Omega_0} M_1(r, \psi, R) d\Omega' = \frac{8\pi}{r} . \tag{A.5.27}$$

It is possible to derive the angular integrals (A.5.23) for the other kernels $M_i(r, \psi, R)$ by means of the recurrence relation (A.5.8). We obtain

$$\int_{\Omega_0} M_{i+1}(r, \psi, R) d\Omega' = - \int_{\Omega_0} \frac{1}{r^{i-2}} \frac{\partial}{\partial r} \left(r^{i-1} M_i(r, \psi, R) \right) d\Omega' =$$

$$= -\frac{1}{r^{i-2}}\frac{\partial}{\partial r}\left(r^{i-1}\int_{\Omega_0} M_i(r,\psi,R)\,d\Omega'\right) . \tag{A.5.28}$$

For $i = 1, 2$ eqn.(A.5.28) reads

$$\int_{\Omega_0} M_2(r,\psi,R)\,d\Omega' = -r\frac{\partial}{\partial r}\left(\int_{\Omega_0} M_1(r,\psi,R)\,d\Omega'\right) = \frac{8\pi}{r} , \tag{A.5.29}$$

$$\int_{\Omega_0} M_3(r,\psi,R)\,d\Omega' = -\frac{\partial}{\partial r}\left(r\int_{\Omega_0} M_2(r,\psi,R)\,d\Omega'\right) = 0 . \tag{A.5.30}$$

Using successively eqn.(A.5.28) we can see that the angular integral (A.5.23) vanishes for all $i > 3$ and we have

$$\int_{\Omega_0} M_i(r,\psi,R)\,d\Omega' = 0 , \qquad\qquad i \geq 3 . \tag{A.5.31}$$

A.5.6 Proofs of eqns.(A.5.11) and (A.5.12)

In this subsection we prove eqns.(A.5.11) and (A.5.12). To do so, we employ mathematical induction, an efficient and simple tool for demonstrating the validity of these relations.

Let us start with eqn.(A.5.11): in the first step of mathematical induction, we show that eqn.(A.5.11) holds for $i = 2$. For $i = 2$, eqn.(A.5.11) reads

$$M_2(r,\psi,R) = r^2\frac{\partial^2 M_0(r,\psi,R)}{\partial r^2} . \tag{A.5.32}$$

Substituting for M_0 from eqn.(A.5.9), we have

$$M_2(r,\psi,R) = -r^2\frac{\partial}{\partial r}\left[\frac{r - R\cos\psi}{\ell^3}\right] = -\frac{r^2}{\ell^3} + \frac{3r^2}{\ell^5}(r - R\cos\psi)^2 . \tag{A.5.33}$$

On the other hand, the kernel $M_2(r,\psi,R)$ may be evaluated from the recurrence formula (A.5.8):

$$M_2(r,\psi,R) = -r\frac{\partial M_1(r,\psi,R)}{\partial r} . \tag{A.5.34}$$

Taking the radial derivative of eqn.(A.5.15) and substituting it into eqn.(A.5.34), we again obtain eqn.(A.5.33). Thus we have demonstrated that eqn.(A.5.11) holds for $i = 2$.

In the next step, we assume that formula (A.5.11) holds for a particular value of i; we wish to prove that it is also valid for $i + 1$. Assuming that eqn.(A.5.11) is valid for M_i, we can substitute it into recurrence formula (A.5.11) obtaining

$$M_{i+1}(r,\psi,R) = \frac{(-1)^{i+1}}{r^{i-2}}\sum_{s=1}^{i-1}\frac{i!(i-2)!}{(i+1-s)!(i-s-1)!(s-1)!}\times$$

$$\times\frac{\partial}{\partial r}\left[r^{2i-s}\frac{\partial^{i+1-s}M_0(r,\psi,R)}{\partial r^{i+1-s}}\right] . \tag{A.5.35}$$

Taking the radial derivative, we have

$$M_{i+1}(r, \psi, R) = (-1)^{i+1} \sum_{s=1}^{i-1} \frac{i!(i-2)!}{(i+1-s)!(i-s-1)!(s-1)!} \times$$

$$\times \left[(2i-s)r^{i+1-s} \frac{\partial^{i+1-s} M_0(r, \psi, R)}{\partial r^{i+1-s}} + r^{i+2-s} \frac{\partial^{i+2-s} M_0(r, \psi, R)}{\partial r^{i+2-s}} \right] . \quad \text{(A.5.36)}$$

Let us now shift the summation index s in the first term to $s+1$. We get:

$$M_{i+1}(r, \psi, R) = (-1)^{i+1} \sum_{s=2}^{i} \frac{i!(i-2)!(2i-s+1)}{(i+2-s)!(i-s)!(s-2)!} r^{i+2-s} \frac{\partial^{i+2-s} M_0(r, \psi, R)}{\partial r^{i+2-s}} +$$

$$+(-1)^{i+1} \sum_{s=1}^{i-1} \frac{i!(i-2)!}{(i+1-s)!(i-s-1)!(s-1)!} r^{i+2-s} \frac{\partial^{i+2-s} M_0(r, \psi, R)}{\partial r^{i+2-s}} . \quad \text{(A.5.37)}$$

Writing separately the contribution with $s = i$ in the first summation and the contribution with $s = 1$ in the second summation, and summing the rest, we get

$$M_{i+1}(r, \psi, R) = (-1)^{i+1} \left[\frac{(i+1)!}{2!} r^2 \frac{\partial^2 M_0(r, \psi, R)}{\partial r^2} + r^{i+1} \frac{\partial^{i+1} M_0(r, \psi, R)}{\partial r^{i+1}} \right] +$$

$$+(-1)^{i+1} \sum_{s=2}^{i-1} \frac{i!(i-2)!}{(i+2-s)!(i-s)!(s-1)!} \times$$

$$\times \left[(s-1)(2i-s+1) + (i-s+2)(i-s) \right] r^{i+2-s} \frac{\partial^{i+2-s} M_0(r, \psi, R)}{\partial r^{i+2-s}} . \quad \text{(A.5.38)}$$

The last expression may be arranged further as follows:

$$M_{i+1}(r, \psi, R) = (-1)^{i+1} \left[\frac{(i+1)!}{2!} r^2 \frac{\partial^2 M_0(r, \psi, R)}{\partial r^2} + r^{i+1} \frac{\partial^{i+1} M_0(r, \psi, R)}{\partial r^{i+1}} \right] +$$

$$+(-1)^{i+1} \sum_{s=2}^{i-1} \frac{(i+1)!(i-1)!}{(i+2-s)!(i-s)!(s-1)!} r^{i+2-s} \frac{\partial^{i+2-s} M_0(r, \psi, R)}{\partial r^{i+2-s}} =$$

$$= (-1)^{i+1} \sum_{s=1}^{i} \frac{(i+1)!(i-1)!}{(i+2-s)!(i-s)!(s-1)!} r^{i+2-s} \frac{\partial^{i+2-s} M_0(r, \psi, R)}{\partial r^{i+2-s}} . \quad \text{(A.5.39)}$$

Finally, comparing eqn.(A.5.39) to eqn.(A.5.11) we can see that eqn.(A.5.39) is equal to eqn.(A.5.11) for subscript $i+1$, and we have thus proved that eqn.(A.5.11) holds also for $M_{i+1}(r, \psi, R)$. Concluding, eqn.(A.5.11) is valid for an arbitrary integer subscript i, Q.E.D.

Now, let us continue with proving eqn.(A.5.12). We will again use mathematical induction. In the first step, we should prove that eqn.(A.5.12) is valid for $k = 1$, but since this is simple, we leave it to the reader. In the next step, we assume that eqn.(A.5.12) holds for subscript k and we show that it is also valid for subscript $k + 1$. Assuming that eqn.(A.5.12) is valid for the k-th radial derivative of the reciprocal distance, we may take another radial derivative of eqn.(A.5.12) getting

$$\frac{\partial^{k+1}}{\partial r^{k+1}}\left(\frac{1}{\ell}\right) = \sum_{t=1}^{k}(-1)^{\frac{k+t}{2}}\frac{(k-t+1)!!(k+t-1)!!}{(k-t+1)!}\frac{k!}{(t-1)!}\frac{(r-R\cos\psi)^{t-1}}{\ell^{k+t+1}} -$$

$$-\sum_{t=0}^{k}(-1)^{\frac{k+t}{2}}\frac{(k-t+1)!!(k+t+1)!!}{(k-t+1)!}\frac{k!}{t!}\frac{(r-R\cos\psi)^{t+1}}{\ell^{k+t+3}} . \qquad (A.5.40)$$

Let us change the summation index t to $t-1$ in the first sum, and to $t+1$ in the second summation. We obtain

$$\frac{\partial^{k+1}}{\partial r^{k+1}}\left(\frac{1}{\ell}\right) = \sum_{t=0}^{k-1}(-1)^{\frac{k+t+1}{2}}\frac{(k-t)!!(k+t)!!}{(k-t)!}\frac{k!}{t!}\frac{(r-R\cos\psi)^{t}}{\ell^{k+t+2}} -$$

$$-\sum_{t=1}^{k+1}(-1)^{\frac{k+t-1}{2}}\frac{(k-t+2)!!(k+t)!!}{(k-t+2)!}\frac{k!}{t-1!}\frac{(r-R\cos\psi)^{t}}{\ell^{k+t+2}} , \qquad (A.5.41)$$

where $k+t+1$ must be an even number. Writing separately the term with $t=0$ in the first summation and the term with $t=k+1$ in the second summation, and summing the rest, we have

$$\frac{\partial^{k+1}}{\partial r^{k+1}}\left(\frac{1}{\ell}\right) = (-1)^{\frac{k+1}{2}}\frac{(k!!)^2}{\ell^{k+2}} - (-1)^k(2k+1)!!\frac{(r-R\cos\psi)^{k+1}}{\ell^{2k+3}} +$$

$$+\sum_{t=1}^{k-1}(-1)^{\frac{k+t+1}{2}}\frac{(k-t+2)!!(k+t)!!}{(k-t+2)!}\frac{k!}{t!}[(k-t+1)+t]\frac{(r-R\cos\psi)^{t}}{\ell^{k+t+2}} . \qquad (A.5.42)$$

By simple manipulations, we finally get

$$\frac{\partial^{k+1}}{\partial r^{k+1}}\left(\frac{1}{\ell}\right) = \sum_{t=0}^{k+1}(-1)^{\frac{k+t+1}{2}}\frac{(k-t+2)!!(k+t)!!}{(k-t+2)!}\frac{(k+1)!}{t!}\frac{(r-R\cos\psi)^{t}}{\ell^{k+t+2}} .$$
$$(A.5.43)$$

Comparing the last equation with eqn.(A.5.12), we can see that we have proved that eqn.(A.5.12) holds also for subscript $k+1$. Therefore, eqn.(A.5.12) is valid for an arbitrary integer number k, Q.E.D.

Chapter 6

The effect of anomalous density of topographical masses

In this chapter we focus our attention on yet another problem, that of the influence of lateral changes of the density of the topographical masses on geoidal height computation. For computing the geoidal heights, the density of the topographical masses is usually modelled by the mean crustal density $\varrho_0 = 2.67$ g/cm^3 (e.g., Vaníček and Kleusberg, 1987; Featherstone, 1992; Sideris, 1994; Forsberg and Sideris, 1993). In the mountains, small variations from this value become significant. Also the density of water in lakes such as the Great Lakes in North America, differs significantly from the value of 2.67 g/cm^3. How much of an error can this cause in geoidal heights?

To answer this question, we express density $\overline{\varrho}(\Omega)$ of topographical masses, assumed to be varying only laterally, as a sum of the constant 'reference' value $\varrho_0 = 2.67$ g/cm^3 and a laterally varying 'anomalous' density $\delta\overline{\varrho}(\Omega)$, i.e.,

$$\overline{\varrho}(\Omega) = \varrho_0 + \delta\overline{\varrho}(\Omega) . \tag{6.1}$$

Substituting the decomposition (6.1) into eqn.(3.30), the density $\sigma(\Omega)$ of the Helmert condensation layer (throughout this chapter we only use Helmert's 2nd condensation technique to reduce the gravitational effect of topographical masses) is written analogously as:

$$\sigma(\Omega) = \sigma_0(\Omega) + \delta\sigma(\Omega) . \tag{6.2}$$

The 'reference' value $\sigma_0(\Omega)$, that corresponds to the reference density ϱ_0, varies only with terrain height:

$$\sigma_0(\Omega) = \varrho_0\,\tau(\Omega) , \tag{6.3}$$

whereas the 'anomalous' condensation density $\delta\sigma(\Omega)$ is associated with both the anomalous density $\delta\overline{\varrho}(\Omega)$ and the height as

$$\delta\sigma(\Omega) = \delta\overline{\varrho}(\Omega)\tau(\Omega) , \tag{6.4}$$

where $\tau(\Omega)$ is given by eqn.(3.62).

6.1 Topographical effects

Now, the direct and both the indirect topographical effects will be specified for the topographical density (6.1) and the condensation density (6.2). Substituting for $\overline{\varrho}(\Omega)$ and $\sigma(\Omega)$ in eqn.(3.45) from eqns.(6.1) and (6.2), the direct topographical effect on gravity can be written as

$$\delta A(\Omega) = \delta A_0(\Omega) + \delta A_{\delta\overline{\varrho}}(\Omega) \,, \tag{6.5}$$

where

$$\delta A_0(\Omega) := \tag{6.6}$$

$$= G\varrho_0 \int_{\Omega_0} \left[\left. \frac{\partial \widetilde{L^{-1}}(r,\psi,r')}{\partial r} \right|_{r'=R+H(\Omega)}^{R+H(\Omega')} - R^2 \left[\tau(\Omega') - \tau(\Omega)\right] \left. \frac{\partial L^{-1}(r,\psi,R)}{\partial r} \right|_{r=R+H(\Omega)} \right] d\Omega' \,,$$

and

$$\delta A_{\delta\overline{\varrho}}(\Omega) := G \int_{\Omega_0} \left[\delta\overline{\varrho}(\Omega') \left. \frac{\partial \widetilde{L^{-1}}(r,\psi,r')}{\partial r} \right|_{r'=R}^{R+H(\Omega')} - \delta\overline{\varrho}(\Omega) \left. \frac{\partial \widetilde{L^{-1}}(r,\psi,r')}{\partial r} \right|_{r'=R}^{R+H(\Omega)} - \right.$$

$$\left. - R^2 \left[\delta\sigma(\Omega') - \delta\sigma(\Omega)\right] \left. \frac{\partial L^{-1}(r,\psi,R)}{\partial r} \right|_{r=R+H(\Omega)} \right] d\Omega' \,. \tag{6.7}$$

Similarly, the primary indirect topographical effect on the geoid $\delta N_{pri}(\Omega)$ caused by the primary indirect topographical effect on potential, $\delta N_{pri}(\Omega) = \delta V(R,\Omega)/\gamma$, where γ is the normal gravity on the level ellipsoid, may be decomposed as

$$\delta N_{pri}(\Omega) = \delta N_{pri,0}(\Omega) + \delta N_{pri,\delta\overline{\varrho}}(\Omega) \,, \tag{6.8}$$

where

$$\delta N_{pri,0}(\Omega) := -\frac{2\pi G}{\gamma}\varrho_0 H^2(\Omega) \left(1 + \frac{2}{3}\frac{H(\Omega)}{R}\right) +$$

$$+\frac{G\varrho_0}{\gamma} \int_{\Omega_0} \left[\left. \widetilde{L^{-1}}(R,\psi,r') \right|_{r'=R+H(\Omega)}^{R+H(\Omega')} - R^2 \frac{\tau(\Omega') - \tau(\Omega)}{L(R,\psi,R)} \right] d\Omega' \,, \tag{6.9}$$

and

$$\delta N_{pri,\delta\overline{\varrho}}(\Omega) := -\frac{2\pi G}{\gamma}\delta\overline{\varrho}(\Omega) H^2(\Omega) \left(1 + \frac{2}{3}\frac{H(\Omega)}{R}\right) +$$

$$+\frac{G}{\gamma} \int_{\Omega_0} \left[\delta\overline{\varrho}(\Omega') \left. \widetilde{L^{-1}}(R,\psi,r') \right|_{r'=R}^{R+H(\Omega')} - \delta\overline{\varrho}(\Omega) \left. \widetilde{L^{-1}}(R,\psi,r') \right|_{r'=R}^{R+H(\Omega)} - \right.$$

$$\left. - R^2 \frac{\delta\sigma(\Omega') - \delta\sigma(\Omega)}{L(R,\psi,R)} \right] d\Omega' \,. \tag{6.10}$$

Finally, the secondary indirect topographical effect on gravity, cf.eqn.(3.8), may be decomposed in a similar fashion:

$$\delta S(\Omega) = \delta S_0(\Omega) + \delta S_{\delta\overline{\varrho}}(\Omega) \,, \tag{6.11}$$

where $\delta S_0(\Omega)$ and $\delta S_{\delta\overline{\varrho}}(\Omega)$ can be expressed by means of respective primary indirect topographical effects on the geoid as

$$\delta S_0(\Omega) := \frac{2\gamma}{R} \delta N_{pri,0}(\Omega) \ , \tag{6.12}$$

and

$$\delta S_{\delta\overline{\varrho}}(\Omega) := \frac{2\gamma}{R} \delta N_{pri,\delta\overline{\varrho}}(\Omega) \ . \tag{6.13}$$

In summary, the terms $\delta A_0(\Omega)$, $\delta N_{pri,0}(\Omega)$ and $\delta S_0(\Omega)$ represent the direct and both the indirect topographical effects induced by topographical masses of constant density ϱ_0, whereas terms $\delta A_{\delta\overline{\varrho}}(\Omega)$, $\delta N_{pri,\delta\overline{\varrho}}(\Omega)$ and $\delta S_{\delta\overline{\varrho}}(\Omega)$ represent the effects induced by column averages of lateral anomalies $\delta\overline{\varrho}(\Omega)$ of the topographical density. The former terms are usually considered in geoid computations (e.g., Vaníček and Kleusberg, 1987, or Sideris, 1994), while the latter terms are not. Since our interest here is focused on exploring the effects of lateral anomalies of topographical density on geoidal heights, we will further investigate only the terms $\delta A_{\delta\overline{\varrho}}(\Omega)$, $\delta N_{pri,\delta\overline{\varrho}}(\Omega)$ and $\delta S_{\delta\overline{\varrho}}(\Omega)$.

6.2   One particular example: a lake

A lake whose surface has a non-zero topographical height represents an obvious example of lateral changes in the topographical density $\overline{\varrho}(\Omega)$ - note that a lake at the altitude of sea level, or ocean, for that matter, has zero topographical height and thus zero topographical effects. We will denote the density of water by ϱ_w, (1.0 g/cm^3), while the density of surrounding topographical masses will be denoted by ϱ_0 (2.67 g/cm^3). Let the orthometric height of the lake surface be $H(\Omega)$ (> 0) and the depth of the lake be $D(\Omega)$, ($D(\Omega) \geq 0$). To a high degree of approximation, we may assume that $H(\Omega) = H_0 = const.$ over the whole lake.

By definition (3.2), the laterally varying density $\overline{\varrho}(\Omega)$ is evaluated from the actual 3-D density $\varrho(r,\Omega)$ by averaging along the topographical column of height $H(\Omega)$. For our example of a lake, this formula yields

$$\overline{\varrho}(\Omega) = \begin{cases} [\varrho_w D(\Omega) + \varrho_0 (H_0 - D(\Omega))] / H_0 \ , & D(\Omega) \leq H_0 \ , \\ \\ \varrho_w \ , & D(\Omega) > H_0 \ . \end{cases} \tag{6.14}$$

The anomalous density $\delta\overline{\varrho}(\Omega)$, cf. eqn.(6.1), is then given by

$$\delta\overline{\varrho}(\Omega) = \begin{cases} (\varrho_0 - \varrho_w)D(\Omega)/H_0 \ , & D(\Omega) \leq H_0 \ , \\ \\ \varrho_0 - \varrho_w \ , & D(\Omega) > H_0 \ . \end{cases} \tag{6.15}$$

The first of eqns.(6.15) shows in particular that if $D(\Omega) = 0$, then $\delta\overline{\varrho}(\Omega) = 0$.

For a lake, the direct topographical effect on gravity given by the term $\delta A_{\delta\overline{\varrho}}(\Omega)$, cf., eqn.(6.7), may be further simplified (Martinec et al., 1995). Realizing that the height of the lake surface is constant, we may put $r' = R + H_0$ (instead of $r' = R + H(\Omega')$) in the first term on the right hand side of eqn.(6.7). Moreover, the second term in this equation is equal to zero when the computation point lies outside the lake because the anomalous density $\delta\bar{\varrho}(\Omega)$ vanishes outside the lake. At the lake, the height $H(\Omega)$ of the computation point is equal to H_0. Therefore, we may put $r' = R + H_0$ (instead of $r' = R + H(\Omega)$) in the second term without changing its numerical value. The term $\delta A_{\delta\bar{\varrho}}(\Omega)$ then becomes

$$\delta A_{\delta\bar{\varrho}}(\Omega) = G \int_{\Omega_0} \left[[\delta\bar{\varrho}(\Omega') - \delta\bar{\varrho}(\Omega)] \frac{\partial \widetilde{L^{-1}}(r,\psi,r')}{\partial r} \Bigg|_{r'=R}^{R+H_0} \right. $$

$$\left. - R^2 [\delta\sigma(\Omega') - \delta\sigma(\Omega)] \frac{\partial L^{-1}(r,\psi,R)}{\partial r} \Bigg|_{r=R+H(\Omega)} \right] d\Omega' . \tag{6.16}$$

Expressing the anomalous condensation densities in the positions Ω and Ω' according to eqn.(6.4), we have

$$\delta\sigma(\Omega) = \delta\bar{\varrho}(\Omega)\,\tau(\Omega) , \qquad \delta\sigma(\Omega') = \delta\bar{\varrho}(\Omega')\,\tau(\Omega') , \tag{6.17}$$

where

$$\tau(\Omega) = H(\Omega)\left(1 + \frac{H(\Omega)}{R} + \frac{H^2(\Omega)}{3R^2}\right) , \qquad \tau(\Omega') = H_0\left(1 + \frac{H_0}{R} + \frac{H_0^2}{3R^2}\right) . \tag{6.18}$$

The condensation density $\delta\sigma(\Omega)$ vanishes if the computation point Ω is outside the lake, therefore, the function $\tau(\Omega)$ may be chosen arbitrarily for this position. At the lake, the function $\tau(\Omega)$ is equal to $\tau(\Omega')$ because of the fixed height of the lake surface. In summary, we may put

$$\tau_0 := \tau(\Omega') = \tau(\Omega) = H_0\left(1 + \frac{H_0}{R} + \frac{H_0^2}{3R^2}\right) . \tag{6.19}$$

The spherical approximation (used for the topographical effects, see sect.3.1) then yields

$$\tau_0 \doteq H_0 . \tag{6.20}$$

Substituting eqns.(6.17) and (6.20) into (6.16), $\delta A_{\delta\bar{\varrho}}(\Omega)$ may be finally written in the following form:

$$\delta A_{\delta\bar{\varrho}}(\Omega) = \tag{6.21}$$

$$= G \int_{\Omega_0} [\delta\bar{\varrho}(\Omega') - \delta\bar{\varrho}(\Omega)] \left[\frac{\partial \widetilde{L^{-1}}(r,\psi,r')}{\partial r} \Bigg|_{r'=R}^{R+H_0} - R^2 H_0 \frac{\partial L^{-1}(r,\psi,R)}{\partial r} \right]_{r=R+H(\Omega)} d\Omega' .$$

The geoidal height increment $\delta N_{pri,\delta\overline{\varrho}}(\Omega)$, cf. eqn.(6.10), may be expressed analogously, getting:

$$\delta N_{pri,\delta\overline{\varrho}}(\Omega) = -\frac{2\pi G}{\gamma}\delta\overline{\varrho}(\Omega)H_0^2\left(1+\frac{2}{3}\frac{H_0}{R}\right)+$$

$$+\frac{G}{\gamma}\int_{\Omega_0}[\delta\overline{\varrho}(\Omega')-\delta\overline{\varrho}(\Omega)]\left[\widetilde{L^{-1}}(R,\psi,r')\Big|_{r'=R}^{R+H_0}-\frac{R^2H_0}{L(R,\psi,R)}\right]d\Omega' . \qquad (6.22)$$

The secondary indirect effect $\delta S_{\delta\overline{\varrho}}(\Omega)$, cf. eqn.(6.13), is simple to derive from the last equation:

$$\delta S_{\delta\overline{\varrho}}(\Omega) = -\frac{4\pi G}{R}\delta\overline{\varrho}(\Omega)H_0^2\left(1+\frac{2}{3}\frac{H_0}{R}\right)+$$

$$+\frac{2G}{R}\int_{\Omega_0}[\delta\overline{\varrho}(\Omega')-\delta\overline{\varrho}(\Omega)]\left[\widetilde{L^{-1}}(R,\psi,r')\Big|_{r'=R}^{R+H_0}-\frac{R^2H_0}{L(R,\psi,R)}\right]d\Omega' . \qquad (6.23)$$

For the highest lake in the world, the lake Titicaca in Peru ($H_0 = 3810$ m), the second term of $\delta N_{pri,\delta\overline{\varrho}}(\Omega)$, the term

$$-\frac{4\pi G}{3\gamma}\delta\overline{\varrho}(\Omega)\frac{H_0^3}{R}$$

has the value of -0.4 mm. Similarly, the corresponding term of $\delta S_{\delta\overline{\varrho}}(\Omega)$, the term

$$-\frac{8\pi G}{3}\delta\overline{\varrho}(\Omega)\frac{H_0^3}{R^2}$$

has the value of -10^{-3} mGals. It is thus not worth considering these terms at all. Neglecting them, eqns.(6.22) and (6.23) read in their final form:

$$\delta N_{pri,\delta\overline{\varrho}}(\Omega) \doteq -\frac{2\pi G}{\gamma}\delta\overline{\varrho}(\Omega)H_0^2+$$

$$+\frac{G}{\gamma}\int_{\Omega_0}[\delta\overline{\varrho}(\Omega')-\delta\overline{\varrho}(\Omega)]\left[\widetilde{L^{-1}}(R,\psi,r')\Big|_{r'=R}^{R+H_0}-\frac{R^2H_0}{L(R,\psi,R)}\right]d\Omega' , \qquad (6.24)$$

and

$$\delta S_{\delta\overline{\varrho}}(\Omega) \doteq -\frac{4\pi G}{R}\delta\overline{\varrho}(\Omega)H_0^2+$$

$$+\frac{2G}{R}\int_{\Omega_0}[\delta\overline{\varrho}(\Omega')-\delta\overline{\varrho}(\Omega)]\left[\widetilde{L^{-1}}(R,\psi,r')\Big|_{r'=R}^{R+H_0}-\frac{R^2H_0}{L(R,\psi,R)}\right]d\Omega' . \qquad (6.25)$$

6.3 Numerical results for the lake Superior

Lake Superior in the central part of North America was chosen to illustrate the effect of lateral changes of topographical density caused by lake water on geoidal heights. The area under study is bounded by latitudes $\phi \doteq 46° - 49°$N and longitudes $\lambda \doteq 268° - 276°$E. Figure 6.1 shows the depth of lake Superior as provided by the Geodetic Survey of Canada. The maximum depth is 329 m, the orthometric height of the lake surface is approximately 183 m.

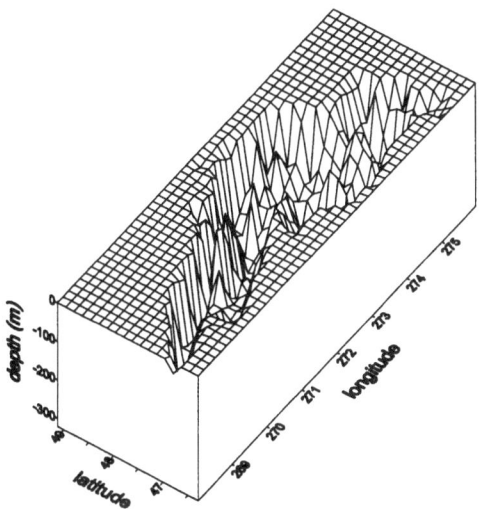

Figure 6.1: Depth of lake Superior (in metres).

Figure 6.2 shows the plot of the term $\delta A_{\delta\overline{\varrho}}(\Omega)$ over the lake Superior. We can see that the magnitude of this term ranges from -0.14 to 0.18 mGal. Stokes's integration of the corrections $\delta A_{\delta\overline{\varrho}}(\Omega)$ provides the increment $\delta N_{dir,\delta\overline{\varrho}}(\Omega)$ to the geoidal heights in the form:

$$\delta N_{dir,\delta\overline{\varrho}}(\Omega) = \frac{R}{4\pi\gamma} \int_{\Omega_0} \delta A_{\delta\overline{\varrho}}(\Omega')S(\psi)d\Omega' . \tag{6.26}$$

Figure 6.3 shows that this increment ranges from -1.1 to 1.3 cm.

To get a better view of the magnitude of the individual terms making the increment $\delta N_{pri,\delta\overline{\varrho}}(\Omega)$ to the geoidal undulations due to the primary indirect topographical effect, let us divide this term into two constituents:

$$\delta N_{pri,\delta\overline{\varrho}}(\Omega) = \delta N^B_{pri,\delta\overline{\varrho}}(\Omega) + \delta N^R_{pri,\delta\overline{\varrho}}(\Omega) , \tag{6.27}$$

where the Bouguer term $\delta N^B_{pri,\delta\overline{\varrho}}(\Omega)$ is equal to

$$\delta N^B_{pri,\delta\overline{\varrho}}(\Omega) := -\frac{2\pi G}{\gamma} \delta\overline{\varrho}(\Omega)H_0^2 . \tag{6.28}$$

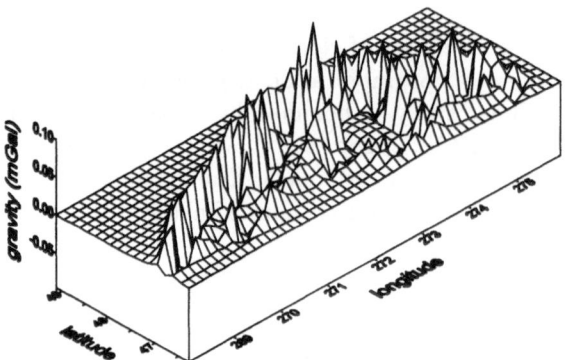

Figure 6.2: The direct topographical effect on gravity $\delta A_{\delta\bar{\varrho}}(\Omega)$ (in mGals) over lake Superior.

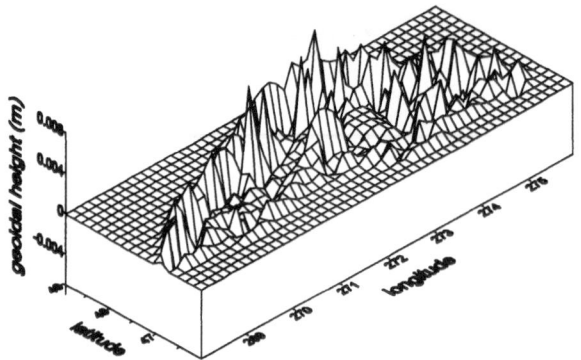

Figure 6.3: The geoid height increment (in metres) generated by corrections $\delta A_{\delta\bar{\varrho}}(\Omega)$ over lake Superior. Truncation radius of Stokes's integration is $6°$.

Figure 6.4 shows the plot of the term $N_{pri,\delta\bar{\varrho}}^{B}(\Omega)$ over lake Superior. We can observe that the magnitude of this term ranges from -0.24 to 0.0 cm, the largest negative values are encountered in the deepest parts of the lake.

Let us explain the reason why the minimal value of -0.24 cm has a 'plateau' over the deepest parts of lake Superior. The orthometric height of the surface of the lake is 183 m, while the depth of the lake reaches the value of 329 m. This means that the deepest water masses of the lake, whose depth is more than 183 m, lie under the geoid. Inspecting equation (6.15) for the anomalous density $\delta\bar{\varrho}(\Omega)$ of the topographical masses, we can find that the "topographical masses" for these deepest parts of the lake have a constant density $\varrho_0 - \bar{\varrho}_w$ (=1.67 g/cm^3). Since the height $H(\Omega)$ of the observer is also constant over the lake, $H(\Omega) = H_0$, eqn.(6.28) shows that the term $\delta N_{pri,\delta\bar{\varrho}}^{B}(\Omega)$ is constant over the deepest parts of the lake, i.e., parts whose the depth is larger than 183 m.

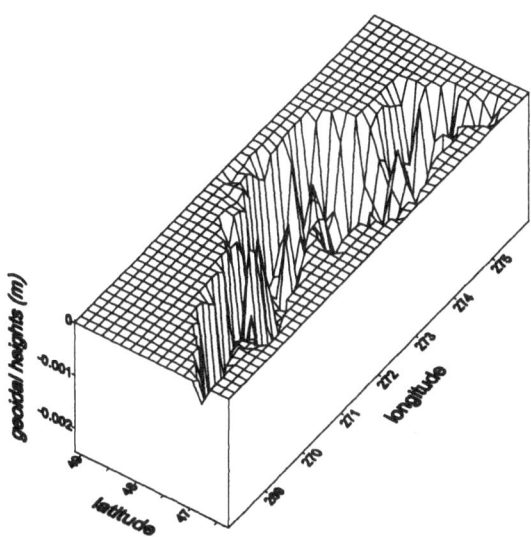

Figure 6.4: The geoid height increment (in metres) generated by the Bouguer part of the primary indirect topographical effect $\delta N^B_{pri,\delta\bar{\varrho}}(\Omega)$ over lake Superior.

The term $\delta N^R_{pri,\delta\bar{\varrho}}(\Omega)$, an analogue of the terrain term in the indirect effect (Heiskanen and Moritz, 1967, sect.3-3) for a case when the density of topographical masses varies laterally, is given by

$$\delta N^R_{pri,\delta\bar{\varrho}}(\Omega) := \frac{G}{\gamma} \int_{\Omega_0} [\delta\bar{\varrho}(\Omega') - \delta\bar{\varrho}(\Omega)] \left[\widetilde{L^{-1}}(R,\psi,r') \Big|_{r'=R}^{R+H_0} - \frac{R^2 H_0}{L(R,\psi,R)} \right] d\Omega' .$$
(6.29)

Figure 6.5 plots the term $\delta N^R_{pri,\delta\bar{\varrho}}(\Omega)$ over lake Superior. The magnitude of this term is of the order of 4×10^{-5} m. Comparing these values with the magnitude of the term $\delta N^B_{pri,\delta\bar{\varrho}}(\Omega)$ (Figure 6.4), we observe that the Bouguer term $\delta N^B_{pri,\delta\bar{\varrho}}(\Omega)$ is about two orders larger than the terrain term $\delta N^R_{pri,\delta\bar{\varrho}}(\Omega)$. This fact can be easily explained by the shape of the bottom of the lake. The slope of the lake banks is very steep so that a larger part of the lake has a depth greater than 183 m. Therefore, replacing real "topographical masses" by the Bouguer plate is a fairly good approximation.

We have also computed the effect of the secondary indirect topographical effect on the geoid, cf., eqn.(6.13). Stokes's integration of $\delta S_{\delta\bar{\varrho}}(\Omega)$ yields the increment $\delta N_{sec,\delta\bar{\varrho}}(\Omega)$. The numerical values of this contribution to the geoid are another of magnitude smaller than the term $\delta N^R_{pri,\delta\bar{\varrho}}(\Omega)$; for a 1 cm geoid both $\delta N^R_{pri,\delta\bar{\varrho}}(\Omega)$ and $\delta N_{sec,\delta\bar{\varrho}}(\Omega)$ may be safely neglected.

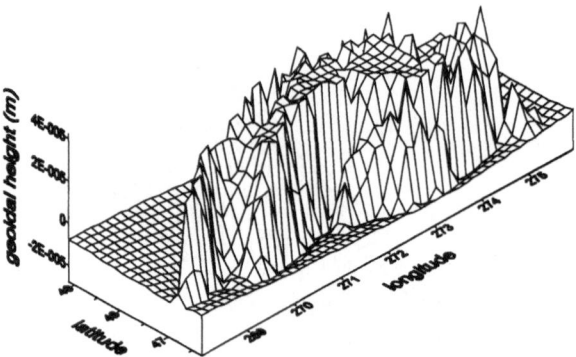

Figure 6.5: The geoid height increment (in metres) generated by the terrain correction to the primary indirect topographical effect $\delta N^R_{pri,\delta\overline{\varrho}}(\Omega)$ over lake Superior.

6.4   Another example: the Purcell Mountains

The largest lateral density variations of topographical masses appear due to a particular geological formation and local geological factors. These may affect the topographical density as much as 10% or even more. In this section we will demonstrate that the lateral changes of topographical densities may play an important role for accurate geoid determination.

For this purpose, we have discretized the Geological Map of Canada (1962) and the Geological Map of British Columbia (1962). We have obtained the data set of rocks and sediments in the region of the Purcell Mountains delimited by latitudes $\phi = 49°N$ and $51°N$, and longitudes $\lambda = 242°E$ and $244°E$ (the eastern part of the Canadian Rocky Mountains). The discretization grids for both the topographical height $H(\Omega)$ and the anomalous density $\delta\overline{\varrho}(\Omega)$ are 5' by 5'. The map of sediments and rocks has then been transformed to a map of anomalous densities $\delta\overline{\varrho}(\Omega)$ according to tables of rock and sediment densities (Telford at al., 1978). The resulting anomalous density of the geological structure beneath the Purcell Mountains is shown in Figure 6.6.

Whereas the first step of this procedure is fairly reliable because there is good knowledge of geological structure of the Rocky Mountains, to associate mass densities to geological rocks or sediments is a highly questionable step because the densities of rocks and sediments are influenced by many factors, e.g., their age, previous history, depth below surface, geological history of the region, etc. The consequence is that the density varies in a large range even for one type of rock or sediment.

The direct topographical effect on gravity $\delta A_{\delta\overline{\varrho}}(\Omega)$ generated by the lateral density variations of geological structure beneath the Purcell Mountains ranges from -6 mGals to 9 mGals (see Figure 6.7). These large contributions to surface

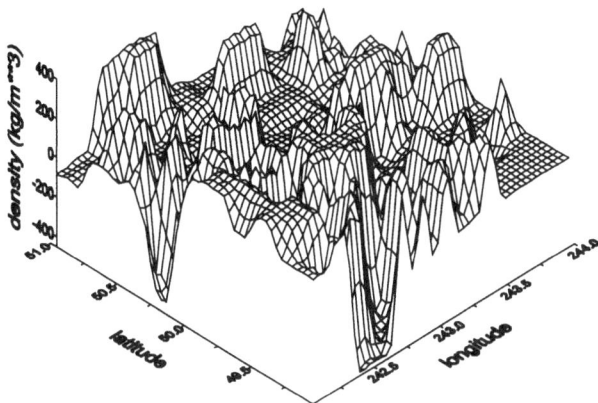

Figure 6.6: The anomalous density $\delta\bar{\varrho}(\Omega)$ (in kg/m³) of the laterally varying geological structure beneath the Purcell Mountains.

gravity anomalies are undoubtedly significant for precise geoid determination, and may contribute by a few decimetres to geoidal heights. Unfortunately, the area plotted in Figure 6.7 is too small that it was impossible to convert the gravity anomalies to the geoidal heights by Stokes's integration.

Figures 6.8 and 6.9 show the Bouguer term $\delta N^B_{pri,\delta\bar{\varrho}}(\Omega)$ and terrain roughness term $\delta N^R_{pri,\delta\bar{\varrho}}(\Omega)$ for the same area of the Purcell Mountains. The values of terms $\delta N^B_{pri,\delta\bar{\varrho}}(\Omega)$ and $N^R_{pri,\delta\bar{\varrho}}(\Omega)$ range between -8 cm and 7 cm, and -0.8 cm and 0.6 cm, respectively. Hence, for 'decimetre geoid' computations, the Bouguer term $\delta N^B_{pri,\delta\bar{\varrho}}(\Omega)$ should be taken into considerations whereas the terrain roughness term $\delta N^R_{pri,\delta\bar{\varrho}}(\Omega)$ may be omitted.

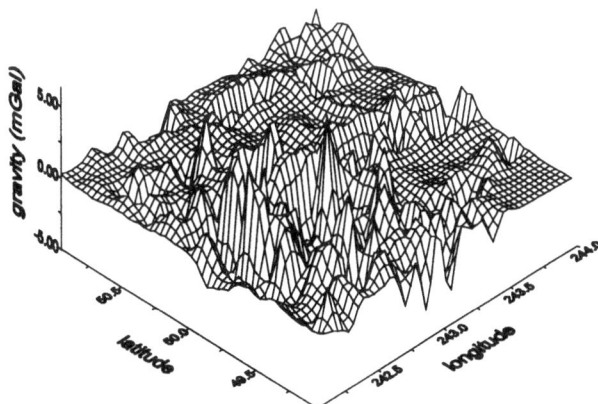

Figure 6.7: The direct topographical effect on gravity $\delta A_{\delta\bar{\varrho}}(\Omega)$ (in mGals) induced by the laterally varying geological structure beneath the Purcell Mountains.

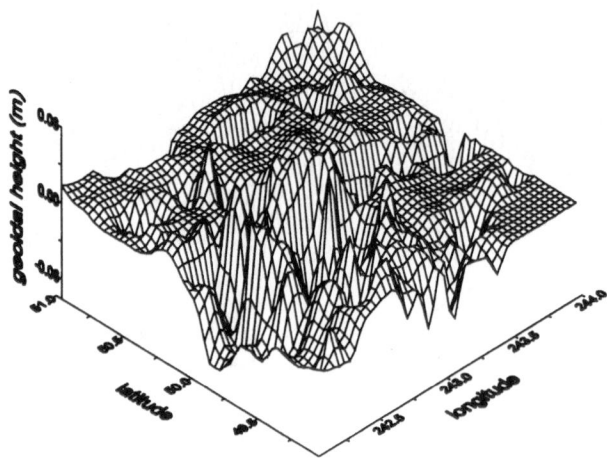

Figure 6.8: The geoidal heights $\delta N^B_{pri,\delta\bar{\varrho}}(\Omega)$ (in metres) induced by the laterally varying geological structure beneath the Purcell Mountains.

6.5   Conclusion

This chapter was motivated by a question whether the lateral density inhomo-geneities of topographical masses should be considered, when the geoid is to be determined with a high accuracy. The standard way of approximating the density of topographical masses when computing their gravitational effects is to take the constant value of 2.67 g/cm^3, while, for instance, the lake water density is 62% less, or local geological factors may change the density of solid topographical masses as much as 10% or even more.

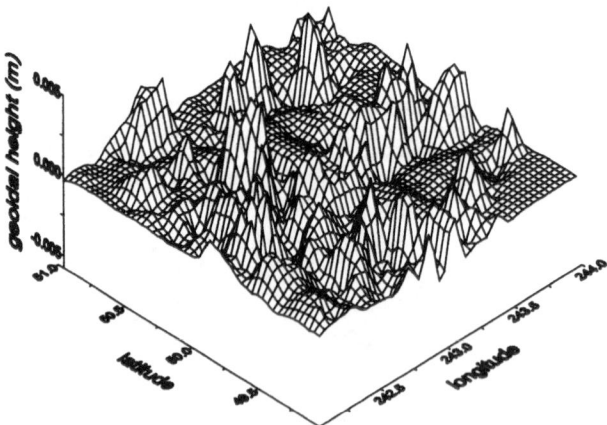

Figure 6.9: The geoidal heights $\delta N^R_{pri,\delta\bar{\varrho}}(\Omega)$ (in metres) induced by the laterally varying geological structure beneath the Purcell Mountains.

Numerical computations have been carried out for lake Superior, the largest of the Great Lakes of North America. We have computed the correction to geoidal heights when the density of 1.0 g/cm^3 was used in the computations instead of the density of 2.67 g/cm^3. The numerical values of these corrections for lake Superior are fairly small; the correction to the geoidal height due to the direct topographical effect on gravity lies within the range (-1.1, 1,3) cm, the correction to the geoidal height due to the primary indirect topographical effect on potential is within (-0.24, 0.0) cm. We have shown that the dominant term in the latter corrections is the Bouguer term (6.28), which depends linearly on the density contrast between water and surrounding rock, and quadratically on the orthometric height of the lake surface.

Moreover, using the Geological Map of Canada we have constructed the map of lateral variations of topographical density for the Purcell Mountains. We have shown that lateral density variations of such a rugged and mountainous terrain may contribute several mGals to the surface gravity, and several decimetres to the geoidal heights. Unfortunately, it is hard to specify these estimates more accurately since lateral density inhomogeneities of the uppermost part of the Earth's crust are still fairly uncertain geophysical quantities.

Chapter 7

Formulation of the Stokes two-boundary-value problem with a higher-degree reference field

In this chapter, our aim is to formulate the Stokes two-boundary-value problem for the determination of the gravimetric geoid considering a higher-degree gravitational potential as a reference. Besides the usual data of surface gravity measurements and heights of the Earth's surface above the geoid, we shall assume that a low-degree potential harmonic expansion obtained from analyses of satellite orbits perturbations, truncated approximately at degree 20, is known a priori, so that it can serve as the reference potential. Let us note that it is the spherical harmonic coefficients of the Earth's gravitational field that are derived from the analysis of satellite orbit perturbations; to obtain the ellipsoidal harmonic coefficients of this field, the transformation relations between spherical and ellipsoidal harmonics derived by Jekeli (1988) are used. Formulating the Stokes two-boundary-value problem for a higher-degree reference potential has several advantages. For instance, the truncation error of Stokes's integral applied to observed gravity data reduced to the reference gravity field is significantly smaller than in the case where the classical Stokes theory is applied to original gravity data that contain low- as well as high-frequency components (Vaníček and Sjöberg, 1991). This reduced truncation error can be evaluated numerically using a global gravity model truncated at degree 120 at most (Martinec, 1993).

However, the formulation of the Stokes two-boundary-value problem with a reference potential given a priori may encounter some difficulties, since it is not as easy as in, for instance, the case of Molodensky's boundary-value problems (Heck, 1991). Whereas Molodensky's problems are governed by the Laplace equation in the external space, and a reference satellite potential represents low-degree components of the solution in the whole space of interest, the reference satellite harmonics used in a partly internal and partly external Stokes two-boundary-value problem represent a solution only in the external space. The gravitational potential induced by topographical masses must be added to the satellite gravity model in order to construct the low-degree part of the solution within the topographical masses.

Vaníček et al. (1995) have made a first attempt to use satellite potential harmonics in the Stokes-Helmert boundary-value problem as the reference. They defined the reference potential for Helmert's anomalous potential as the difference between the satellite model and the low-degree components of the direct topographical effect on the potential. This means that the reference harmonics of the sought potential depend on the way the topographical masses are condensed. In this chapter, we will show that the reference potential for the Stokes two-boundary-value problem can be introduced differently, such that it does not depend on the way the topographical masses are compensated or condensed. This model better reflects the physical and mathematical background of the this boundary-value problem, because the long-wavelength part of its solution is uniquely determined by the boundary conditions on the geoid and the Earth's surface and, of course, by the Laplace-Poisson equation. Another question not answered by Vaníček et al. (1995) is how to reduce the observed surface gravity to the reference field. Again, we will show that such a reduction can be performed without specifying the model of compensation of topographical masses.

Let us first formulate the Stokes two-boundary-value problem described by eqns.(1.46)–(1.48) and (1.24) in terms of the anomalous gravitational potential T instead of the potential T^h. Moreover, let this free boundary-value problem be treated only as a fixed one. To present the theory as transparent as possible, we shall approximate the geoid by a mean sphere of radius R, $r_g(\Omega) = R$. There is, however, no principal difficulty to introduce ellipsoidal approximation of the geoid when a low-degree reference field is to be introduced (Martinec and Vaníček, 1996). The decomposition (1.14) in formulation (1.46)–(1.48) yields

$$\nabla^2 T = -4\pi G\varrho \qquad \text{outside } S_g , \qquad (7.1)$$

$$\left.\frac{\partial T}{\partial r}\right|_{r=R+H(\Omega)} + \left.\frac{2}{R}T\right|_{r=R} - \epsilon_h(T)|_{r=R+H(\Omega)} - \epsilon_\gamma(T)|_{r=R} = -\Delta g^F , \qquad (7.2)$$

$$T \sim O\left(\frac{1}{r^3}\right) \qquad r \to \infty . \qquad (7.3)$$

The symbols used have the same meaning as in Chapter 1.

7.1 A higher-degree reference gravitational potential

Let us assume that some low-degree harmonics of the gravitational potential T have been determined from satellite orbit analyses. The question arises of how to reformulate the fixed boundary-value problem (7.1)–(7.3) so that a low-degree satellite gravity model can be considered as a reference gravitational potential.

Since the anomalous gravitational potential T is harmonic outside the Earth and it vanishes at infinity, it can be represented as a series of solid spherical harmonics $r^{-j-1}Y_{jm}(\Omega)$, $j = 2, 3, ...$, and $|m| \leq j$, which all vanish at infinity (Heiskanen and Moritz, 1967, eqn.(1-46)):

$$T(r, \Omega) = \sum_{j=2}^{\infty} \left(\frac{R}{r}\right)^{j+1} \sum_{m=-j}^{j} T_{jm} Y_{jm}(\Omega) . \tag{7.4}$$

This series is convergent outside the bounding sphere $r = R_1$ completely enclosing the Earth, but it may be divergent in the space between the Earth's surface and this bounding sphere. To define a higher-degree reference potential for the Stokes two-boundary-value problem, however, we are interested only in the low-degree part of the potential T. Let us thus split the potential T into the (known) low-degree reference potential T_ℓ and a (unknown) higher-degree gravitational potential T^ℓ:

$$T = T_\ell + T^\ell . \tag{7.5}$$

The set of potential coefficients T_{jm}^e (where superscript e stands for external) determined by analyses of perturbations of satellite orbits will be taken as the reference. The reference gravitational potential T_ℓ in the space outside the bounding sphere $r = R_1$ is then represented as

$$T_\ell(r, \Omega) = \sum_{j=2}^{\ell} \left(\frac{R}{r}\right)^{j+1} \sum_{m=-j}^{j} T_{jm}^e Y_{jm}(\Omega) , \tag{7.6}$$

where ℓ is a cut-off degree of the reference potential coefficients, say $\ell = 20$. Moreover, since $|T_{jm}^e| < \infty$, the finite series (7.6) has finite values not only outside the bounding sphere $r = R_1$, but also in the space between the Earth's surface and the sphere $r = R_1$. Therefore, T_ℓ represented by the finite series (7.6) can be considered as the reference gravitational potential for gravity observations performed on the Earth's surface.

Analogously, the residual topographical potential δV can also be split into low-degree and high-degree parts:

$$\delta V = \delta V_\ell + \delta V^\ell . \tag{7.7}$$

The low-degree part δV_ℓ is given by low-degree parts of expansions (A.7.12) derived in Appendix A.7:

$$\delta V_\ell(r, \Omega) = \begin{cases} \sum_{j=0}^{\ell} \left(\frac{R}{r}\right)^{j+1} \sum_{m=-j}^{j} \left(V_{jm}^{t,e} - V_{jm}^c\right) Y_{jm}(\Omega) & \text{for } r \geq R + H(\Omega) , \\ \sum_{j=0}^{\ell} \sum_{m=-j}^{j} \left(V_{jm}^{t,i} - V_{jm}^c\right) Y_{jm}(\Omega) & \text{for } r = R . \end{cases}$$
$$\tag{7.8}$$

Applying the same argument as in the preceeding paragraph, we represent δV_ℓ by the first series not only above $r = R_1$ but also between the Earth's surface and the bounding sphere $r = R_1$. Note that formulae (7.8) do not determine the potential $\delta V_\ell(r, \Omega)$ within the topographical masses.

7.2 Reference gravity anomaly

The crucial point of the problem of a higher-degree reference field is determining the (low-frequency) part Δg_ℓ^F of the free-air gravity anomaly Δg^F that is generated by the reference potential T_ℓ. Obviously, Δg_ℓ^F is given by boundary condition (7.2) applied to T_ℓ,

$$\Delta g_\ell^F := -\left.\frac{\partial T_\ell}{\partial r}\right|_{r=R+H(\Omega)} - \left.\frac{2}{R} T_\ell\right|_{r=R} + \epsilon_h(T_\ell)|_{r=R+H(\Omega)} + \epsilon_\gamma(T_\ell)|_{r=R} . \quad (7.9)$$

The first and third terms of Δg_ℓ^F can easily be computed employing the representation (7.6) of T_ℓ. Unfortunately, the same representation cannot be used for evaluating the second and fourth terms of Δg_ℓ^F, because formula (7.6) is valid only outside the Earth (not on the geoid). Hence, the next step will be devoted to deriving spherical harmonic expansion of T_ℓ at a point on the geoid.

The gravity potential W can be considered to be a sum of gravitational potential V^g generated by the masses below the geoid, the topographical potential V^t, and the centrifugal potential V^ω:

$$W = V^g + V^t + V^\omega. \quad (7.10)$$

The two different decompositions of gravity potential W, eqn.(7.10) and $W = U + T$, can now be put together so that the potential T reads

$$T = V^t + V^g + V^\omega - U . \quad (7.11)$$

This equation is valid everywhere inside and outside the Earth. Outside the geoid, in particular, the gravitational potential $V^g + V^\omega - U$ is harmonic, and it can be represented by an spherical harmonic series of the form (valid also for $r = R$)

$$V^g + V^\omega - U = \sum_{j=0}^{\infty} \left(\frac{R}{r}\right)^{j+1} \sum_{m=-j}^{j} (V^g + V^\omega - U)_{jm} Y_{jm}(\Omega) \qquad \text{for } r \geq R , \quad (7.12)$$

where $(V^g + V^\omega - U)_{jm}$ are expansion coefficients. Substituting the last formula together with the expansions (A.7.5) and (7.6) into eqn.(7.11), comparing the coefficients by solid spherical harmonics $r^{-j-1}Y_{jm}(\Omega)$ up to degree ℓ, and considering the continuation property of harmonic functions, in particular the unique extension of the region of definition of a harmonic function (Kellogg, 1953, sect.10.5), we obtain

$$(V^g + V^\omega - U)_{jm} = T_{jm}^e - V_{jm}^{t,e} \qquad \text{for } r \geq R , \quad (7.13)$$

where $j = 0, 1, ..., \ell$, $|m| \leq j$, and $V_{jm}^{t,e}$ are given by eqn.(A.7.6).

On the geoid, $r = R$, the formula (7.11) together with the expansions (A.7.7) and (7.12), yields

$$T(R, \Omega) = \sum_{j=0}^{\infty} \sum_{m=-j}^{j} \left(V_{jm}^{t,i} + (V^g + V^\omega - U)_{jm}\right) Y_{jm}(\Omega) . \quad (7.14)$$

The low-degree part T_ℓ of potential T, see decomposition (7.5), at a point on the geoid then reads

$$T_\ell(R, \Omega) = \sum_{j=0}^{\ell} \sum_{m=-j}^{j} \left(V_{jm}^{t,i} + (V^g + V^\omega - U)_{jm} \right) Y_{jm}(\Omega) , \qquad (7.15)$$

or, on substituting for $(V^g + V^\omega - U)_{jm}$ from eqn.(7.13), we have

$$\boxed{T_\ell(R, \Omega) = \sum_{j=0}^{\ell} \sum_{m=-j}^{j} T_{jm}^i Y_{jm}(\Omega) ,} \qquad (7.16)$$

where

$$\boxed{T_{jm}^i := T_{jm}^e + V_{jm}^{t,i} - V_{jm}^{t,e} ,} \qquad (7.17)$$

$j = 0, 1, ..., \ell$, and $|m| \leq j$.

Finally, we are ready to evaluate the reference free-air gravity anomaly Δg_ℓ^F. Substituting eqns.(7.6) and (7.16) into (7.9), we obtain

$$\boxed{\begin{aligned} \Delta g_\ell^F(\Omega) = \frac{1}{R} \sum_{j=0}^{\ell} \sum_{m=-j}^{j} \left[(j+1) \left(\frac{R}{R + H(\Omega)} \right)^{j+2} T_{jm}^e - 2T_{jm}^i \right] Y_{jm}(\Omega) + \\ + \epsilon_h(T_{jm}^e) + \epsilon_\gamma(T_{jm}^i) . \end{aligned}} \qquad (7.18)$$

It is important that the reference free-air gravity anomaly Δg_ℓ^F does not depend on the way the topographical masses are compensated or condensed, but only on the reference satellite harmonics T_{jm}^e, the density distribution of topographical masses via differences $V_{jm}^{t,i} - V_{jm}^{t,e}$, and on topographical height $H(\Omega)$.

7.3 Formulation of the two-boundary-value problem

Adding eqn.(7.9) to (7.2) and using the decomposition (7.5), we have

$$\left. \frac{\partial T^\ell}{\partial r} \right|_{r=R+H(\Omega)} + \left. \frac{2}{R} T^\ell \right|_{r=R} - \epsilon_h(T^\ell) \Big|_{r=R+H(\Omega)} - \epsilon_\gamma(T^\ell) \Big|_{r=R} = -\Delta g^{F,\ell} , \quad (7.19)$$

where T^ℓ is the high-degree part of the potential T, and $\Delta g^{F,\ell}$ is the high-degree part of free-air gravity anomaly,

$$\Delta g^{F,\ell} := \Delta g^F - \Delta g_\ell^F . \qquad (7.20)$$

On the strength of the starting assumption of spherical approximation of the geoid, the residual topographical potential δV, and thus also its high-degree part

δV^{ℓ}, can be considered as known quantities at points on the Earth's surface and the geoid. The latter quantity can readily be determined from formula (7.7), where δV_{ℓ} is given by spherical harmonic expansion (7.8). This makes it possible to introduce a new unknown potential $T^{h,\ell}$:

$$T^{h,\ell} := T^{\ell} - \delta V^{\ell} . \tag{7.21}$$

By noting that the function $T - \delta V$ is harmonic outside the geoid, its high-degree part $T^{h,\ell}$, i.e., the high-degree part of Helmert's so-called anomalous potential (Martinec et al., 1993) when the topographical masses are compensated according to Helmert's second condensation technique, satisfies the boundary-value problem of the form

$$\nabla^2 T^{h,\ell} = 0 \qquad\qquad r > R , \tag{7.22}$$

$$\left.\frac{\partial T^{h,\ell}}{\partial r}\right|_{r=R+H(\Omega)} + \frac{2}{R}T^{h,\ell}\Big|_{r=R} - \epsilon_h(T^{h,\ell})\Big|_{r=R+H(\Omega)} - \epsilon_\gamma(T^{h,\ell})\Big|_{r=R} =$$
$$= -\Delta g^{F,\ell} - \delta A^{\ell} - \delta S^{\ell} + \epsilon_h(\delta V^{\ell})\Big|_{r=R+H(\Omega)} + \epsilon_\gamma(\delta V^{\ell})\Big|_{r=R} , \tag{7.23}$$

$$T^{h,\ell} \sim O\left(\frac{1}{r^{\ell+1}}\right) \qquad\qquad r \to \infty , \tag{7.24}$$

where boundary condition (7.23) follows immediately from substitution of eqn. (7.21) into (7.19). The high-degree parts of the direct and secondary indirect topographical effects on gravity read

$$\delta A^{\ell} := \left.\frac{\partial \delta V(r,\Omega)}{\partial r}\right|_{R+H(\Omega)} + \frac{1}{R}\sum_{j=0}^{\ell}(j+1)\left(\frac{R}{R+H(\Omega)}\right)^{j+2}\sum_{m=-j}^{j}\left(V_{jm}^{t,e} - V_{jm}^{c}\right)Y_{jm}(\Omega) \tag{7.25}$$

and

$$\delta S^{\ell} := \frac{2}{R}\delta V(R,\Omega) - \frac{2}{R}\sum_{j=0}^{\ell}\sum_{m=-j}^{j}\left(V_{jm}^{t,i} - V_{jm}^{c}\right)Y_{jm}(\Omega) . \tag{7.26}$$

The asymptotic constrain (7.24) reflects condition (7.3) and the fact that only the high-degree part of the gravitational potential is sought, while the low-degree component T_{ℓ} is assumed to be completely known a priori. Not even the best satellite gravity model is error-free, however, leaving an unmodelled low-degree residual in the boundary data. The following question may arise: should we try to determine this low-frequency residual or should we be satisfied with the low-frequency part of the solution as defined by a satellite gravity model and a low-degree model of topography, and only search for the high-frequency part of the solution? Nowadays, we tend to accept the latter point of view, since the coverage of gravimetric data over the Earth's surface is still not accurate, dense and homogeneous enough to be used to determine low-frequency components of the Earth's gravity field more precisely. This concept is also supported by the fact that the modelling the long-wavelength term δV_{ℓ} of the residual topographical potential is not error-free because of an insufficient knowledge of the density of topographical masses. Thus, some long-wavelength residual of the gravitational effect of topographical masses affects the surface gravity data. Without knowledge of the 3-D density structure of topographical masses, it is impossible to distinguish this residual from long-wavelength errors of a satellite gravity model.

Accepting this concept, i.e., assuming that the solution is sought only for high-degree components (even though the reduced boundary data contain some low-frequency noise), the solution to the problem (7.22)–(7.24) may be affected by a low-degree aliasing effect. Whether or not the high-frequency part of a solution T^ℓ we are looking for is distorted by low-frequency noise in the boundary data depends on how the problem is solved. For instance, if the above boundary-value problem is solved by means of Poisson's and Stokes's integrals, the integration kernels must be constructed such that they are 'blind' to low-frequency components of the boundary data. Then the low-degree harmonics do not affect the high-frequency solution. The construction of Poisson's and Stokes's integration kernels with such a desired property can be found in Vaníček et al. (1987; 1996).

It is not the intention of this chapter to construct the solution to the problem (7.22)–(7.24). For a long time, geodesists have been solving more or less similar problems. In most cases, they have tried to transform geodetic boundary-value problems similar to the problem (7.22)–(7.24) to Stokes's problem and to solve them by Stokes's integration (Heiskanen and Moritz, 1967, sect.2-26). However, a satisfactory solution to the problem (7.22)–(7.24) matching today's accuracy requirements has not yet been presented. Perhaps the most crucial problems are that (1) the existence of a solution to the problem (7.22)–(7.24) cannot be guaranteed (see Chapter 1), and (2) the solution is unstable (see Chapter 8). What conditions guarantee the existence of the solution and how to reasonably stabilize the solution are the open questions which have not yet been answered satisfactorily.

After finding $T^{h,\ell}$ by solving the fixed boundary-value problem (7.22)–(7.24), the solution to the free boundary-value problem (1.1)–(1.4) can be found by giving the undulations N of the geoid with respect to the reference ellipsoid. Substituting eqns.(1.14), (7.5) and (7.21) into Bruns's formula (1.24), we obtain

$$N = N_\ell + \frac{1}{\gamma_Q} \left(T^{h,\ell} + \delta V^\ell \right) \Big|_{P_g} , \qquad (7.27)$$

where we have introduced low-degree geoidal undulations N_ℓ as

$$N_\ell := \frac{1}{\gamma_Q} T_\ell \Big|_{P_g} . \qquad (7.28)$$

From eqns.(7.16) and (7.17), we obtain

$$N_\ell = \frac{1}{\gamma_Q} \sum_{j=0}^{\ell} \sum_{m=-j}^{j} \left(T_{jm}^e + V_{jm}^{t,i} - V_{jm}^{t,e} \right) Y_{jm}(\Omega) . \qquad (7.29)$$

It should be emphasized that neither Δg_ℓ^F nor N_ℓ depends on the way the topographical masses are compensated or condensed, but only on the reference harmonics T_{jm}^e and on the differences $V_{jm}^{t,i} - V_{jm}^{t,e}$ of spherical harmonics induced by topographical masses. On the other hand, the boundary-value problem (7.22)–(7.24) for $T^{h,\ell}$ depends on the way the topographical masses are compensated.

7.4   Numerical results for $V_{jm}^{t,i} - V_{jm}^{t,e}$

Let us try now to estimate the effect of the gravitational field generated by topographical masses on the reference free-air gravity anomaly Δg_ℓ^F and the reference geoidal undulations N_ℓ. To do this, we shall compute the differences $V_{jm}^{t,i} - V_{jm}^{t,e}$ for low degrees, $j = 0, 1, .., 20$. As a first approximation of $V_{jm}^{t,i} - V_{jm}^{t,e}$, we shall assume that the density $\varrho(r, \Omega)$ of topographical masses is constant and equal to the mean crustal density $\varrho_0 = 2.67$ g/cm^3. The actual density of topographical masses is expected to vary around ϱ_0 by 10 to 20 per cent. Later on, we will estimate the effect of such topographical density variations on the differences $V_{jm}^{t,i} - V_{jm}^{t,e}$.

Taking the density of topographical masses constant, formula (A.7.6) for $V_{jm}^{t,e}$ reduces to

$$V_{jm}^{t,e} = \frac{4\pi G \varrho_0 R^{-j-1}}{2j+1} \int_{\Omega_0} \int_{r'=R}^{R+H'} r'^{j+2} Y_{jm}^*(\Omega') dr' d\Omega' , \tag{7.30}$$

where we have denoted $H' := H(\Omega')$ to abbreviate following notations. Carrying out the integration over r', we obtain

$$V_{jm}^{t,e} = \frac{4\pi G \varrho_0 R^{-j-1}}{(2j+1)(j+3)} \int_{\Omega_0} \left[(R+H')^{j+3} - R^{j+3} \right] Y_{jm}^*(\Omega') d\Omega' . \tag{7.31}$$

Expanding the term in square brackets according to the binomial theorem yields

$$\frac{1}{j+3} \left[(R+H')^{j+3} - R^{j+3} \right] = R^{j+3} \sum_{i=1}^{j+3} \frac{1}{i} \binom{j+2}{i-1} \left(\frac{H'}{R} \right)^i . \tag{7.32}$$

The potential coefficients of topographical potential now read

$$V_{jm}^{t,e} = \frac{4\pi G \varrho_0 R^2}{2j+1} \sum_{i=1}^{j+3} \frac{1}{i} \binom{j+2}{i-1} \int_{\Omega_0} \left(\frac{H'}{R} \right)^i Y_{jm}^*(\Omega') d\Omega' . \tag{7.33}$$

The second set of potential coefficients $V_{jm}^{t,i}$, defined by eqn.(A.7.8), may be arranged analogously:

$$V_{jm}^{t,i} = \frac{4\pi G \varrho_0 R^j}{2j+1} \int_{\Omega_0} \int_{r'=R}^{R+H'} r'^{-j+1} Y_{jm}^*(\Omega') dr' d\Omega' , \tag{7.34}$$

where the definite integral over r' may take two values:

$$\int_{r'=R}^{R+H'} (r')^{-j+1} dr' = \begin{cases} \frac{-1}{-j+2} [(R+H')^{-j+2} - R^{-j+2}] & \text{if } j \neq 2 \\ \\ \ln(1 + H'/R) & \text{if } j = 2 . \end{cases} \tag{7.35}$$

Making use of the binomial series expansion, we have

$$\frac{1}{-j+2}\left[(R+H')^{-j+2} - R^{-j+2}\right] = R^{-j+2}\sum_{i=1}^{\infty}\frac{1}{i}\left(\begin{array}{c}-j+1\\i-1\end{array}\right)\left(\frac{H'}{R}\right)^i, \quad (7.36)$$

for $j \neq 2$. Once $j = 2$, the binomial coefficient takes the value ($\forall\, i \geq 1$):

$$\left(\begin{array}{c}-j+1\\i-1\end{array}\right) = \left(\begin{array}{c}-1\\i-1\end{array}\right) = (-1)^{i-1}. \quad (7.37)$$

The function $\ln(1 + H'/R)$ in eqn.(7.35) can be developed into Taylor's series expansion as follows

$$\ln\left(1 + \frac{H'}{R}\right) = \sum_{i=1}^{\infty}\frac{(-1)^{i-1}}{i}\left(\frac{H'}{R}\right)^i. \quad (7.38)$$

The comparison of eqn.(7.38) with eqn.(7.36) for $j = 2$ shows that the series expansion on the right-hand side of eqn.(7.36) for $j = 2$ is of the same form as that in eqn.(7.36). Thus, the primitive function to integral (7.35) can be written in a unified form as

$$\int_{r'=R}^{R+H'} (r')^{-j+1}dr' = R^{-j+2}\sum_{i=1}^{\infty}\frac{1}{i}\left(\begin{array}{c}-j+1\\i-1\end{array}\right)\left(\frac{H'}{R}\right)^i. \quad (7.39)$$

The potential coefficients (7.34) now read

$$V_{jm}^{t,i} = \frac{4\pi G\varrho_0 R^2}{2j+1}\sum_{i=1}^{\infty}\frac{1}{i}\left(\begin{array}{c}-j+1\\i-1\end{array}\right)\int_{\Omega_0}\left(\frac{H'}{R}\right)^i Y_{jm}^*(\Omega')d\Omega'. \quad (7.40)$$

Finally, we are ready to determine the differences $V_{jm}^{t,i} - V_{jm}^{t,e}$. Subtracting eqn.(7.31) from (7.40) yields

$$V_{jm}^{t,i} - V_{jm}^{t,e} = \frac{4\pi G\varrho_0 R^2}{2j+1}\sum_{i=1}^{\infty}\frac{1}{i}\left[\left(\begin{array}{c}-j+1\\i-1\end{array}\right) - \left(\begin{array}{c}j+2\\i-1\end{array}\right)\right]\int_{\Omega_0}\left(\frac{H'}{R}\right)^i Y_{jm}^*(\Omega')d\Omega'. \quad (7.41)$$

It is important that the contribution with $i = 1$ is equal to zero because the factor in square brackets vanishes. Writing down explicitly the first three non-zero contributions in eqn.(7.41), we obtain

$$V_{jm}^{t,i} - V_{jm}^{t,e} = -2\pi G\varrho_0\left[\left(H^2\right)_{jm} + \frac{2}{3R}\left(H^3\right)_{jm} + \frac{j(j+1)}{12R^2}\left(H^4\right)_{jm} + \ldots\right], \quad (7.42)$$

where $\left(H^i\right)_{jm}$ are spherical harmonic coefficients of the i-th power of H,

$$\left(H^i\right)_{jm} = \int_{\Omega_0}[H(\Omega)]^i\, Y_{jm}^*(\Omega)d\Omega. \quad (7.43)$$

To determine the reference free-air gravity anomaly Δg_ℓ^F and the reference geoidal undulations N_ℓ, only low-degree harmonics $V_{jm}^{t,i} - V_{jm}^{t,e}$ are to be computed. In this case, the spherical harmonics $(H^i)_{jm}$ for $i = 4, 5, \ldots$ may safely be neglected, and eqn.(7.42) becomes

$$V_{jm}^{t,i} - V_{jm}^{t,e} \doteq -2\pi G\varrho_0 \left[\left(H^2\right)_{jm} + \frac{2}{3R} \left(H^3\right)_{jm} \right] . \tag{7.44}$$

Table 7.1 gives the differences $V_{jm}^{t,i} - V_{jm}^{t,e}$, $j = 0, 1, \ldots, 20$, $|m| \leq j$, for the TUG87 global spherical harmonic terrain model (Wieser, 1987) complete up to degree and order 180. Spherical harmonic coefficients $(H^i)_{jm}$ of the i-th power of the topographical heights have been computed numerically according to the approach proposed by Martinec (1989). We have found that contributions of $V_{jm}^{t,i} - V_{jm}^{t,e}$, $j = 0, 1, \ldots, 20$, $|m| \leq j$, to geoidal heights lie within the interval (-2.80; 0) m; the minimum -2.80 m is located in the Himalayas. Note that this result is in an agreement with that obtained by Sjöberg (1994). A plot of this effect for the territory of Canada is shown in Figure 7.1. As expected, the minimum value, which reaches -0.43 m, is connected with the highest part of the Canadian Rocky Mountains. Since the density of topographical masses enters Newton's integral linearly, in order to achieve the 1 dm accuracy in geoidal heights, the regional density of the Rocky Mountains massif should be known with a relative accuracy of better than 25 per cent.

Inspecting eqn.(7.18), we can observe that the part of the reference free-air gravity anomaly $\Delta g_\ell^F(\Omega)$ that originates from differences $V_{jm}^{t,i} - V_{jm}^{t,e}$ is

$$\Delta g_{\ell,topo}^F(\Omega) \doteq -\frac{2}{b_0} \sum_{j=0}^{\ell} \sum_{m=-j}^{j} \left(V_{jm}^{t,i} - V_{jm}^{t,e} \right) . \tag{7.45}$$

We have evaluated $\Delta g_{\ell,topo}^F(\Omega)$ globally for $\ell = 20$ and found that it reaches at most 0.09 mGal.

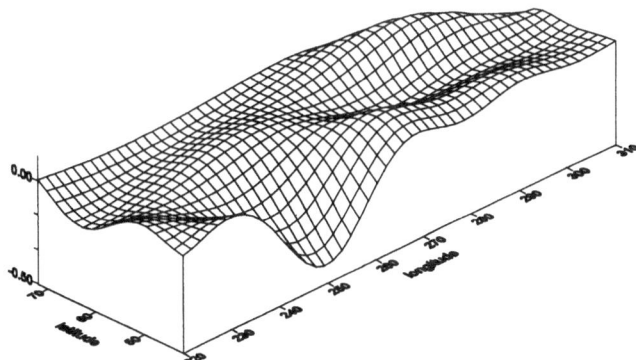

Figure 7.1: The part of the reference geoidal heights N_ℓ (in metres) that originates from differences $V_{jm}^{t,i} - V_{jm}^{t,e}$ over the territory of Canada.

j	m	$V^{t,i}_{jm} - V^{t,e}_{jm}$		j	m	$V^{t,i}_{jm} - V^{t,e}_{jm}$		j	m	$V^{t,i}_{jm} - V^{t,e}_{jm}$	
0	0	-0.51	0.00	8	4	0.09	0.03	12	2	-0.04	0.02
1	0	0.02	0.00	8	5	-0.02	-0.18	12	3	0.13	0.15
1	1	-0.10	-0.23	8	6	0.15	-0.04	12	4	-0.09	0.02
2	0	-0.32	0.00	8	7	-0.09	0.10	12	5	0.03	0.05
2	1	0.02	-0.24	8	8	0.02	0.00	12	6	-0.15	0.04
2	2	0.27	-0.04	9	0	0.25	0.00	12	7	0.01	-0.02
3	0	0.48	0.00	9	1	-0.12	-0.40	12	8	-0.07	0.02
3	1	0.00	-0.23	9	2	0.08	0.04	12	9	-0.02	-0.15
3	2	0.32	-0.09	9	3	-0.01	-0.10	12	10	0.04	-0.04
3	3	0.04	0.06	9	4	0.22	-0.03	12	11	0.03	-0.02
4	0	-0.18	0.00	9	5	0.05	0.01	12	12	0.02	0.02
4	1	0.01	0.37	9	6	0.12	-0.04	13	0	-0.16	0.00
4	2	0.27	-0.02	9	7	-0.02	0.22	13	1	-0.07	-0.13
4	3	-0.06	0.30	9	8	-0.18	0.03	13	2	-0.23	0.08
4	4	-0.13	-0.09	9	9	0.00	-0.07	13	3	-0.03	-0.03
5	0	0.42	0.00	10	0	0.07	0.00	13	4	-0.08	0.06
5	1	-0.07	0.07	10	1	0.05	0.27	13	5	0.00	-0.09
5	2	-0.01	0.05	10	2	0.24	-0.02	13	6	-0.04	0.02
5	3	-0.03	0.26	10	3	0.06	0.12	13	7	0.00	-0.11
5	4	-0.26	0.14	10	4	0.17	-0.09	13	8	0.07	0.02
5	5	-0.01	-0.14	10	5	0.03	0.12	13	9	-0.05	-0.06
6	0	-0.47	0.00	10	6	0.03	-0.03	13	10	0.06	0.02
6	1	0.05	0.39	10	7	0.01	0.18	13	11	-0.04	0.11
6	2	-0.21	0.05	10	8	-0.08	0.02	13	12	-0.03	-0.08
6	3	0.02	0.05	10	9	-0.02	-0.04	13	13	0.04	-0.04
6	4	-0.26	0.07	10	10	-0.04	0.00	14	0	0.00	0.00
6	5	0.08	-0.16	11	0	0.15	0.00	14	1	0.06	-0.06
6	6	0.09	0.03	11	1	-0.11	0.02	14	2	-0.03	0.03
7	0	0.05	0.00	11	2	0.02	0.04	14	3	0.05	-0.09
7	1	-0.14	-0.25	11	3	-0.01	0.21	14	4	0.02	0.02
7	2	-0.31	0.07	11	4	0.02	0.02	14	5	-0.04	-0.14
7	3	-0.06	-0.11	11	5	0.01	0.12	14	6	0.05	0.01
7	4	-0.12	0.00	11	6	-0.12	0.01	14	7	-0.03	-0.12
7	5	-0.01	-0.25	11	7	0.00	0.06	14	8	0.11	-0.03
7	6	0.18	-0.06	11	8	-0.10	0.01	14	9	-0.02	-0.02
7	7	0.05	0.12	11	9	0.03	-0.19	14	10	0.04	0.02
8	0	-0.24	0.00	11	10	0.08	-0.01	14	11	0.03	0.09
8	1	0.02	0.07	11	11	-0.03	0.06	14	12	-0.01	0.06
8	2	0.00	0.08	12	0	-0.02	0.00	14	13	-0.04	-0.02
8	3	0.00	-0.17	12	1	0.10	0.30	14	14	0.00	-0.03

Table 7.1. Spherical harmonic coefficients $V^{t,i}_{jm} - V^{t,e}_{jm}$ (in m^{2}s^{-2}) for the TUG87 global terrain model (Wieser, 1987) and for $\varrho_0 = 2.67$ g/cm^3.

15	0	-0.01	0.00		17	4	0.06	0.00		19	4	-0.07	0.07
15	1	-0.06	-0.20		17	5	0.05	0.08		19	5	0.05	-0.05
15	2	0.05	0.08		17	6	0.01	-0.07		19	6	-0.06	0.00
15	3	-0.08	-0.10		17	7	0.03	0.07		19	7	-0.01	-0.07
15	4	0.13	0.06		17	8	-0.07	-0.03		19	8	-0.05	0.04
15	5	-0.01	-0.06		17	9	0.05	0.02		19	9	-0.02	-0.06
15	6	0.10	-0.01		17	10	-0.05	0.03		19	10	0.04	0.01
15	7	-0.03	0.00		17	11	0.03	-0.05		19	11	-0.03	-0.04
15	8	0.06	-0.04		17	12	0.00	0.00		19	12	0.04	-0.03
15	9	-0.02	0.06		17	13	-0.04	-0.03		19	13	0.01	0.02
15	10	-0.04	-0.02		17	14	-0.02	0.04		19	14	0.03	0.02
15	11	0.03	0.04		17	15	-0.05	-0.03		19	15	0.01	0.00
15	12	-0.02	-0.04		17	16	0.05	-0.02		19	16	0.06	-0.02
15	13	0.06	0.00		17	17	0.02	0.03		19	17	-0.03	0.05
15	14	0.00	0.06		18	0	0.02	0.00		19	18	-0.05	-0.01
15	15	-0.02	0.00		18	1	0.04	0.12		19	19	-0.01	-0.01
16	0	0.17	0.00		18	2	-0.02	-0.06		20	0	-0.06	0.00
16	1	0.06	-0.02		18	3	0.11	0.16		20	1	0.01	-0.13
16	2	0.17	-0.06		18	4	-0.06	-0.04		20	2	-0.08	-0.05
16	3	0.09	0.07		18	5	0.05	0.06		20	3	0.05	-0.08
16	4	0.15	-0.06		18	6	-0.09	-0.03		20	4	-0.04	0.00
16	5	0.03	0.05		18	7	0.03	-0.01		20	5	0.03	-0.09
16	6	0.10	-0.04		18	8	-0.11	0.00		20	6	0.03	0.01
16	7	0.01	0.08		18	9	0.02	-0.06		20	7	-0.01	-0.11
16	8	0.00	-0.03		18	10	-0.01	0.01		20	8	0.03	0.02
16	9	0.01	0.10		18	11	0.01	-0.07		20	9	-0.02	-0.02
16	10	-0.07	0.01		18	12	0.06	-0.03		20	10	0.08	0.02
16	11	0.00	0.01		18	13	0.02	0.02		20	11	-0.01	0.03
16	12	-0.04	-0.04		18	14	0.01	0.04		20	12	0.02	0.01
16	13	-0.04	-0.06		18	15	0.02	0.05		20	13	-0.01	0.05
16	14	0.02	-0.02		18	16	-0.06	0.01		20	14	-0.02	0.04
16	15	0.05	0.05		18	17	-0.02	-0.04		20	15	-0.01	-0.03
16	16	0.00	0.03		18	18	0.00	0.00		20	16	0.00	-0.04
17	0	0.02	0.00		19	0	-0.10	0.00		20	17	0.01	-0.04
17	1	-0.01	0.12		19	1	0.01	0.03		20	18	0.03	0.02
17	2	0.06	0.02		19	2	-0.12	0.04		20	19	-0.01	0.01
17	3	-0.04	0.12		19	3	-0.03	-0.01		20	20	-0.02	0.00

Table 7.1. (Continued.)

7.5 Conclusion

To employ a reference potential in the computation of the gravimetric geoid has an evident advantage; it reduces the magnitude of quantities we work with and thus enables us to linearize the originally non-linear boundary-value problem for geoid determination. There is certainly a lot of possible ways to bring the reference potential into the geoid computation. Here, a reference potential has been considered to consist of the harmonics derived from the analyses of satellite orbits. We have assumed that such a priori information on long-wavelength components of the gravitational field is fixed and should not be corrected from surface gravity observations. This fact is expressed by an asymptotic condition (7.24), which says that only the short-wavelength part of the gravitational potential is sought from surface gravity data. On the other hand, introducing a reference potential into the problem of geoid determination requires that the reference potential harmonics are accurate enough and that they contain meaningful information. Therefore, we have assumed that reference satellite harmonics are taken up to degree about 20.

It should be emphasized that satellite potential harmonics define the reference gravitational potential in the space external to the Earth. In order to construct a reference potential for a partly internal and partly external boundary-value problem of geoid determination, the low-degree part of the gravitational potential induced by topographical masses must be taken into account. Vaníček et al. (1995) have already formulated the boundary-value problem for geoid determination for the case where a satellite reference potential is taken as the reference. They confined themselves to the so-called Stokes-Helmert technique for geoid computation, and introduced the reference potential for Helmert's anomalous potential as a satellite gravitational potential minus the direct topographical effect on the potential. Evidently, such a reference field depends on the way the topographical masses are condensed.

Here, we were motivated by whether the Stokes two-boundary-value problem could be formulated in such a way that a higher-degree reference potential would be independent of the way the topographical masses are compensated. We have shown that such a formulation exists; the reference free-air gravity anomaly as well as the reference geoidal height are determined by a satellite gravitational model and by the differences of the external and internal gravitational fields generated by topographical masses. The reference potential for geoid determination is insensitive to the way the topographical masses are compensated. Numerically, the magnitude of that part of the reference geoidal heights which comes from low-degree components of the topographical potential is approximately three times larger than the corresponding direct topographical effect on the potential in the Stokes-Helmert technique (Vaníček et al., 1995).

A.7 Spherical harmonic representation of δV

In this Appendix, we express the residual topographical potential δV in terms of spherical harmonics. The expansion of the reciprocal distance in terms of the Legendre polynomials $P_j(\cos\psi)$ can be found in Kellogg (1929, sect.5.2); for

$r > r'$, it holds

$$\frac{1}{L(r, \psi, r')} = \frac{1}{r} \sum_{j=0}^{\infty} \left(\frac{r'}{r} \right)^j P_j(\cos \psi) . \qquad (A.7.1)$$

Using the Laplace addition theorem for spherical harmonics (e.g., Varshalovich et al., 1989, sect.5.17.2),

$$P_j(\cos \psi) = \frac{4\pi}{2j+1} \sum_{m=-j}^{j} Y_{jm}^*(\Omega') Y_{jm}(\Omega) , \qquad (A.7.2)$$

the reciprocal distance may further be written as

$$\frac{1}{L(r, \psi, r')} = \frac{4\pi}{r} \sum_{j=0}^{\infty} \frac{1}{2j+1} \left(\frac{r'}{r} \right)^j \sum_{m=-j}^{j} Y_{jm}^*(\Omega') Y_{jm}(\Omega) . \qquad (A.7.3)$$

Let us now turn our attention to the topographical potential V^t. Substituting the above expansion of the reciprocal distance into Newton's integral (1.5), the gravitational potential V^t at a point outside the sphere (of minor semi-axis R_1, say) completely enclosing the Earth reads

$$V^t(r, \Omega) = \frac{4\pi G}{r} \int_{\Omega_0} \int_{r'=r_g(\Omega')}^{r_g(\Omega')+H(\Omega')} \varrho(r', \Omega') \sum_{j=0}^{\infty} \frac{1}{2j+1} \left(\frac{r'}{r} \right)^j \times$$

$$\times \sum_{m=-j}^{j} Y_{jm}^*(\Omega') Y_{jm}(\Omega) r'^2 dr' d\Omega' , \qquad (A.7.4)$$

For $r \geq R_1$, it is admissible to interchange the order of summation over j and m with the integration over r' and Ω' because of the uniform convergence of the series. We get

$$V^t(r, \Omega) = \sum_{j=0}^{\infty} \left(\frac{R}{r} \right)^{j+1} \sum_{m=-j}^{j} V_{jm}^{t,e} Y_{jm}(\Omega) \qquad \text{for } r \geq R_1 , \qquad (A.7.5)$$

where the expansion coefficients $V_{jm}^{t,e}$ read

$$V_{jm}^{t,e} := \frac{4\pi G R^{-j-1}}{2j+1} \int_{\Omega_0} \int_{r'=R}^{R+H(\Omega')} \varrho(r', \Omega') r'^{j+2} Y_{jm}^*(\Omega') dr' d\Omega' . \qquad (A.7.6)$$

Here, we have introduced factors $1/R^{j+1}$ to normalize the expansion coefficients $V_{jm}^{t,e}$.

Analogously, for $r \leq R$, the gravitational potential V^t induced by the topographical masses may be expressed in terms of solid spherical harmonics that are regular at the origin, i.e., as

$$V^t(r, \Omega) = \sum_{j=0}^{\infty} \left(\frac{r}{R} \right)^j \sum_{m=-j}^{j} V_{jm}^{t,i} Y_{jm}(\Omega) , \qquad \text{for } r \leq R , \qquad (A.7.7)$$

with coefficients

$$V_{jm}^{t,i} := \frac{4\pi G R^j}{2j+1} \int_{\Omega_0} \int_{r'=R}^{R+H(\Omega')} \varrho(r',\Omega') r'^{-j+1} Y_{jm}^*(\Omega') dr' d\Omega' . \qquad (A.7.8)$$

As we have discussed, the compensation of topographical masses plays an important role in geoid determination. Employing expansion (A.7.3) of the Newton kernel, the compensation potential V^c, eqn.(1.9), at points outside/on the geoid may be expressed as a series of solid spherical harmonics:

$$V^c(r,\Omega) = \sum_{j=0}^{\infty} \left(\frac{R}{r}\right)^{j+1} \sum_{m=-j}^{j} V_{jm}^c Y_{jm}(\Omega) \qquad \text{for } r \geq R , \qquad (A.7.9)$$

where the coefficients V_{jm}^c are equal to

$$V_{jm}^c := \frac{4\pi G R^{-j-1}}{2j+1} \int_{\Omega_0} \int_{r'=r_c(\Omega')}^{R} \varrho_c(r',\Omega') r'^{j+2} Y_{jm}^*(\Omega') dr' d\Omega' \qquad (A.7.10)$$

for the isostatic compensation of topographical masses, and

$$V_{jm}^c := \frac{4\pi G R}{2j+1} \int_{\Omega_0} \sigma(\Omega') Y_{jm}^*(\Omega') dr' d\Omega' \qquad (A.7.11)$$

for Helmert's second condensation.

Finally, we are ready to write the spherical harmonic representation of the residual topographical potential $\delta V = V^t - V^c$. Using expansions (A.7.5), (A.7.7), and (A.7.9), the potential δV reads

$$\delta V(r,\Omega) = \begin{cases} \displaystyle\sum_{j=0}^{\infty} \left(\frac{R}{r}\right)^{j+1} \sum_{m=-j}^{j} \left(V_{jm}^{t,e} - V_{jm}^c\right) Y_{jm}(\Omega) & \text{for } r \geq R_1 , \\[2em] \displaystyle\sum_{j=0}^{\infty} \sum_{m=-j}^{j} \left(V_{jm}^{t,i} - V_{jm}^c\right) Y_{jm}(\Omega) & \text{for } r = R , \end{cases}$$

$$(A.7.12)$$

where the coefficients $V_{jm}^{t,e}$, $V_{jm}^{t,i}$, and V_{jm}^c are given by eqns.(A.7.6), (A.7.8), and (A.7.10), respectively.

Chapter 8

A discrete downward continuation problem for geoid determination

The problem of downward continuation of the gravity field from the Earth's surface to the geoid arises from the fact that the solution to the boundary-value problem for geoid determination is sought in terms of the gravitational potential on the geoid but the gravity observations are only available on the Earth's surface. Unfortunately, the Earth's surface for continental areas differs significantly from the geoid and thus the geoid potential parameters must be derived from surface gravity functionals smoothed or damped to some extent. The downward continuation of gravity field not only from the Earth's surface to the geoid but also from the satellite or aerial altitudes to the Earth's surface is one of the trickiest problem of physical geodesy since it belongs to the group of the ill-posed problems. An extended list of papers in geodetic literature studying this problem during last three decades documentates it quite obviously: Pellinen (1962), Bjerhammar (1962, 1963, 1976, 1987), Sjöberg (1975), Ilk (1987, 1993), Moritz (1980a, sect.45), Rummel (1979), Cruz (1985), Wang (1988, 1990), Engels et al. (1993), Vaníček at al. (1996), among others. Despite of some advancements, a further effort would be necessary to solve still open questions related to this problem.

The downward continuation of gravity data is an ill-posed problem in the sense that the results (on gravitational potential) do not continuously depend on the observations. The consequence is that errors contaminated data together with roundoff errors result in high-frequency oscillations in the solution. To get a stable solution, some regularization approach is to be applied. A common technique of regularization in context of geoid determination, used, e.g., by Sideris and Forsberg (1990) or Wang (1988, 1990), has been suggested by by Moritz (1980a, sect.45). In this approach, the Taylor series expansion transforms the gravity anomalies from the Earth's surface to the geoid (see sect.1.8). The Taylor series expansion is usually approximated by the first term g_1 only (Moritz,

1980a, sect.47). To avoid the problem with the ill-posedness, the term g_1 is not evaluated properly; it is assumed that the gravity anomalies are linearly dependent on topographical heights and the integration for the term g_1 is taken over topographical heights and not over the surface gravity data. However, the linear relationship between free-air gravity anomalies and topographical heights introduced by Pellinen (1962) holds only approximately (Heiskanen and Moritz, 1967, Figure 7-6). The open question remains to answer how large are the errors of geoidal heights due to adopting this approximation.

The downward continuation problem of gravitational field is governed by Fredholm's integral equation of the 1st kind, symbolically written as

$$Ku = f \, , \tag{8.1}$$

where K is an integral operator, f is a known function, and function u is to be determined. To regularize the solution of this equation, Ilk (1987, 1993), or Engels et al. (1993), suggested to apply Tikhonov's regularization method to eqn.(1), and solve Fredholm's integral equation of the 2nd kind,

$$(K + \alpha L)u = f \, , \tag{8.2}$$

where L is a regularization operator and α is a regularization parameter. The choice of an optimal regularization operator L and optimal regularization parameter α is the main problem of Tikhonov's regularization technique. For instance, if L is the identity operator, and α is too large, the error due to regularization will be too large, and the solution will be too smooth. The consequence is that information contained in observations will be lost adequately. On the contrary, if α is chosen too small, data errors will be amplified too large, and the norm of an approximate solution will be extremely large.

Bjerhammar (1962, 1963, 1976, 1987) was one of the first geodesists who began to deal with the discrete downward continuation problem with claiming that gravity observations do not cover the Earth's surface continuously, but, in practical applications, they are known in discrete points only. He discretized Poisson's integral postulating that gravity anomalies are harmonic in space between the Earth's surface and a fully embedded sphere. The established system of linear equations was solved for gravity anomalies on the embedded sphere. Cruz (1985) applied his technique, called the Dirac approach, for continuing the surface $2^o \times 2^o$ and $5' \times 5'$ gravity anomalies (from the area of New Mexico) down to the embedded sphere. He demonstrated that iterations of the Dirac approach converge faster than the least squares collocation method. However, this result need not necessarily mean that the Dirac approach was well-posed in those particular cases.

In Vaníček et al. (1996) we have independently proposed a very similar technique to the Dirac approach for finding the solution of the downward continuation problem occurring in geoid height determination problem. Removing the gravitational effect of topographical masses from the surface gravity data by Helmert's condensation technique, the gravitational field in space above the geoid becomes harmonic, and the relationship between Helmert's gravity field on the Earth's surface and the geoid is expressed by Poisson's integral. That is why, we discretized Poisson's kernel in a regular grid $5' \times 5'$ for which mean topographical heights and mean surface gravity anomalies were to our disposal, and solved a large system of linear algebraic equations for gravity data selected from a rugged terrain of the Canadian Rocky Mountains. We did not use any kind of regularization other than integral averaging over geographical cells.

In this chapter, we are aiming to investigate the stability of the discrete downward continuation problem in a rigorous way. We particularly intend to give an answer to the following problem. An unstable character of the continuous downward continuation problem is stabilized to some extent by a spatial discretization of a solution because a high frequency part of a solution, determination of which makes the problem unstable, is a priori excluded from a solution. The cut-off frequency is given by the Nyquist frequency associated with a spatial grid step size. A question arises on the smallest spatial grid step size for which a solution to the discrete downward continuation problem still has a stable solution.

We shall proceed in the following way (see also Martinec, 1996). We again employ the discretized form of Poisson's integral to set up the system of linear algebraic equations describing the problem. In contrast to Vaníček et al. (1996) we do not modify Poisson kernel but assume that the contribution of far-zone integration points to the Poisson integral is determined to a sufficient accuracy using one of the existing global gravitational model of the Earth. The eigenvalue spectrum of the matrix of system of linear equations and the spectral contents of surface gravity anomalies are two items controlling the stability of the discrete downward continuation problem. We will carry out the eigenvalue analysis of the matrix of system of equations and analyse the extremal eigenvalues to express the conditionality of this matrix. The conditionality of the matrix will depend on the elevation of the Earth's topography above the geoid since the topographical heights enter the Poisson kernel. We will see that there is a principal difficulty in determining the minimum eigenvalue of this matrix since it is related to the spectral norm of the inverse matrix we will not be able to construct.

The compensation of topographical masses is another possible way how to stabilize the problem as the spectral contents of the gravity anomalies of compensated topographical masses may significantly differ from that of the original free-air gravity anomalies. Using surface observables from the Canadian Rocky Mountains, we will investigate the efficiency of highly idealized compensation models to dampen high-frequency oscillations of the free-air gravity anomalies. The maximum entropy spectrum method will help us to analyze the power spectral contents of the surface gravity observations.

Having analysed the stability of the downward continuation problem, we suggest simple Jacobi's iterative scheme for solving linear equations. This enables to find a solution to the problem without storing the matrix of system of equations which indeed created some problems in our initial attempt (Vaníček et al, 1996). Finally, the convergency of Jacobi's iterations is checked by the spectral norm of

the matrix of linear system of equations which uniquely controls the convergency
of the iterations.

8.1   Formulation of the boundary-value problem

To begin with, let us recall the formulation of the Stokes two-boundary-value
problem for geoid determination with a higher-degree reference gravity field. This
problem leads to find an anomalous gravitational potential $T^{h,\ell}(r, \Omega)$ that satisfies
the following linearized boundary-value problem

$$\nabla^2 T^{h,\ell} \;=\; 0 \qquad\qquad \text{for } r > r_g(\Omega) , \quad (8.3)$$

$$\left.\frac{\partial T^{h,\ell}}{\partial r}\right|_{r_g(\Omega)+H(\Omega)} + \left.\frac{2}{r_g} T^{h,\ell}\right|_{r_g(\Omega)} \;=\; -\Delta g^{h,\ell} , \qquad\qquad (8.4)$$

$$T^{h,\ell} \;\sim\; O\left(\frac{1}{r^{\ell+1}}\right) \qquad r \to \infty , \qquad\quad (8.5)$$

where the geoid S_g is described by an angularly dependent function $r = r_g(\Omega)$
and function $H = H(\Omega)$ is the height of the Earth's surface above the geoid
reckoned along the geocentric radius. Since we intend to solve the problem of
downward continuation of the gravity field from the Earth's surface to the geoid,
we shall assume that $H(\Omega) \geq H_{min} > 0$ throughout this chapter. Once $H = 0$,
the downward continuation problem has a trivial solution.

Asymptotic condition (8.5) states (which is also emphasized by the second
superscript ℓ at $T^{h,\ell}$) that only a high-frequency part of the gravitational potential
$T^{h,\ell}$ is looked for by solving the problem (8.3)–(8.5). We thus assume that low-
degree components of gravitational field are prescribed a priori as the reference
field. For instance, a satellite gravitational model cut approximately at degree
$\ell = 20$ can be taken as the reference. The right-hand sides of the boundary
condition (8.4), gravity anomalies $\Delta g^{h,\ell}$, are assumed to be given continuously
on the Earth's surface. They consist of high-frequency parts $\Delta g^{F,\ell}(\Omega)$, $\delta A^{\ell}(\Omega)$,
and $\delta S^{\ell}(\Omega)$ of the free-air gravity anomalies, the direct topographical effect on
gravity and the secondary indirect topographical effect on gravity, respectively:

$$\Delta g^{h,\ell}(\Omega) := \Delta g^{F,\ell}(\Omega) + \delta A^{\ell}(\Omega) + \delta S^{\ell}(\Omega) . \qquad\qquad (8.6)$$

We refer the reader to Chapter 7 for the definitions of these quantities.

Three approximations will be imposed on the boundary-value problem (8.3)–
(8.5). First, we have already omitted the ellipsoidal corrections $\epsilon_\gamma(T^h)$ and $\epsilon_h(T^h)$
occurring in the original formulation (7.22)–(7.24) of the Stokes two-boundary-
value problem since they have no noticeable contributions to potential T^h beyond
degree $\ell = 20$ (Cruz, 1986). Second, we will treat the problem (8.3)–(8.5) as a
fixed boundary-value problem taking the spherical approximation of the geoid
into account. This means that the radius of the geoid in the potential $T^{h,\ell}$ will

be approximated by a mean radius of the Earth, i.e., $r_g(\Omega) \doteq R$. This spherical approximation of the geoid guarantees an accuracy better then 3×10^{-3}. Since the downward continuation of surface gravity anomalies contributes to geoidal heights by 1 metre at most (Vaníček et al., 1996), the error of the downward continuation problem due to spherical approximation of the geoid does not exceed a 1 cm level. This approximation is thus admissible in most practical applications and will be used throughout this chapter. Third, we will refer the second term on the left-hand side of eqn.(8.4) to the Earth's surface (see the discussion after eqn.(1.67)).

Under these approximations, the problem (8.3)–(8.5) may be reformulated in terms of function $\tau^\ell(r, \Omega)$,

$$\tau^\ell(r, \Omega) := r \frac{\partial T^{h,\ell}}{\partial r} + 2T^{h,\ell} , \tag{8.7}$$

as

$$\nabla^2 \tau^\ell = 0 \qquad \text{for } r > R , \tag{8.8}$$

$$\tau^\ell \big|_{R+H(\Omega)} = f(\Omega) , \tag{8.9}$$

$$\tau^\ell \sim O\left(\frac{1}{r^{\ell+1}}\right) \qquad r \to \infty , \tag{8.10}$$

where $f(\Omega)$ stands for a known 'data' functional prescribed on the Earth's surface,

$$f(\Omega) := -[R + H(\Omega)]\Delta g^{h,\ell}(\Omega) . \tag{8.11}$$

Our goal is to solve boundary-value problem (8.8)–(8.10) and find function τ^ℓ in space outside the geoid. Particularly, we are looking for τ^ℓ on the geoid, i.e., $\tau^\ell(R, \Omega)$. The last problem is often called the <u>downward continuation of a harmonic function</u> (in our case, function τ^ℓ) since a harmonic function is computed on the lower boundary (the geoid) from its value on the upper boundary (the Earth's surface) (Heiskanen and Moritz, 1967, sect.8-10).

8.2 Poisson's integral

Let us first turn our attention to the external spherical Dirichlet problem for the Laplace equation,

$$\nabla^2 \tau^\ell = 0 \qquad \text{for } r > R , \tag{8.12}$$

$$\tau^\ell \big|_R = g(\Omega) , \tag{8.13}$$

$$\tau^\ell \sim O\left(\frac{1}{r^{\ell+1}}\right) \qquad r \to \infty , \tag{8.14}$$

where g is assumed to be a known square-integrable function. This problem may be thought as opposite to the problem (8.8)–(8.10) since function τ^ℓ is looked for outside the geoid $(r = R)$ from its values $g(\Omega)$ on the geoid. In geodetic literature this problem is called the upward continuation of a harmonic function (Heiskanen and Moritz, 1967, sect.6-6).

By contrast to problem (8.8)–(8.10), the solution to the upward continuation problem can be found by a simple integration. Writing the harmonic representation of function $\tau^\ell(r, \Omega)$, $r > R$, as

$$\tau^\ell(r, \Omega) = \sum_{j=l}^{\infty} \sum_{m=-j}^{j} \left(\frac{R}{r}\right)^{j+1} \tau_{jm}^\ell Y_{jm}(\Omega) , \tag{8.15}$$

where $r^{-j-1} Y_{jm}(\Omega)$ are complex solid spherical harmonics and τ_{jm}^ℓ are expansion coefficients to be determined, substituting expansion (8.15) into boundary condition (8.13), expanding function $g(\Omega)$ in a series of spherical harmonics, and comparing the coefficients at spherical harmonics in the resulting formula, we get

$$\tau_{jm}^\ell = \int_{\Omega_0} g(\Omega) Y_{jm}^*(\Omega) d\Omega . \tag{8.16}$$

By coefficients τ_{jm}^ℓ and the Laplace addition theorem for spherical harmonics (e.g., Varshalovich et al., 1989, sect.5.17.2), the solution to upward continuation problem may be expressed in terms of the well-known Poisson's integral (Kellogg, 1929, sect.9.4)

$$\tau^\ell(r, \Omega) = \frac{1}{4\pi} \int_{\Omega_0} \tau^\ell(R, \Omega') K^\ell(r, \psi, R) d\Omega' \tag{8.17}$$

with the kernel

$$K^\ell(r, \psi, R) := \sum_{j=\ell}^{\infty} (2j + 1) \left(\frac{R}{r}\right)^{j+1} P_j(\cos \psi) , \tag{8.18}$$

where ψ is the angular distance between the geocentric directions Ω and Ω', and $P_j(\cos \psi)$ is Legendre polynomial of degree j. By the analogy with terminology introduced for Stokes's functions (Vaníček and Kleusberg, 1987), we will call $K^\ell(r, \psi, R)$ the spheroidal Poisson's kernel. This can easily be expressed by means of the spherical Poisson kernel $K(r, \psi, R)$ as

$$K^\ell(r, \psi, R) = K(r, \psi, R) - \sum_{j=0}^{\ell-1} (2j + 1) \left(\frac{R}{r}\right)^{j+1} P_j(\cos \psi) , \tag{8.19}$$

where (Kellogg, 1929, sect.9.4)

$$K(r, \psi, R) := R \frac{r^2 - R^2}{L^3(r, \psi, R)} , \tag{8.20}$$

and $L(r, \psi, R)$ stands for spatial distance between points (r, Ω) and (R, Ω').

The integration in Poisson's integral (8.17) is to be taken over the full solid angle. For regional geoid determination, it is advantageous to divide the integration domain Ω_0 into near- and far-zone integration subdomains. The near-zone subdomain is created by a spherical cap (of a small radius ψ_0) surrounding the computation point while the rest of the full solid angle forms the far-zone subdomain. The radius ψ_0 of the near-zone spherical cap may be chosen by various manners; one choice is introduced in Appendix A.8 by eqn.(A.8.4). In Appendix B.8, we derive the contributions of particular integration sub-domains to Poisson's integral (8.17). Schematically, Poisson's integral (8.17) may be written as a sum of three terms having different origins,

$$\tau^\ell(r, \Omega) = \tau_0^\ell(r, \Omega) + \tau_{\psi_0}^\ell(r, \Omega) + \tau_{\pi - \psi_0}^\ell(r, \Omega) , \tag{8.21}$$

where $\tau_0^\ell(r, \Omega)$ expresses the contribution to Poisson's integral from the integration point being on the same geocentric radius as the computation point, eqn.(B.8.16), $\tau_{\psi_0}^\ell(r, \Omega)$ expresses the contributions of integration points lying within the near-zone spherical cap of radius ψ_0 (except the point $\Omega' = \Omega$), eqn.(B.8.8), and $\tau_{\pi - \psi_0}^\ell(r, \Omega)$ expresses the contribution of far-zone integration points, eqn.(B.8.28). As a matter of fact, Poisson's kernel $K^\ell(r, \psi, R)$ decreases rapidly with growing angular distance ψ. The dominant behaviour of Poisson's kernel in the vicinity of point $\psi = 0$ implies that $\tau_0^\ell(r, \Omega)$ and $\tau_{\psi_0}^\ell(r, \Omega)$ reach much larger values than $\tau_{\pi - \psi_0}^\ell(r, \Omega)$. For instance, in the Canadian Rocky Mountains, $\tau_0^\ell(r, \Omega)$ and $\tau_{\psi_0}^\ell(r, \Omega)$ reach several tens of mGal while $\tau_{\pi - \psi_0}^\ell(r, \Omega)$ only a few hundreds of μGal (Vaníček et al., 1996). Later on, this fact will be utilized for excluding far-zone contribution $\tau_{\pi - \psi_0}^\ell(r, \Omega)$ from unknowns to be determined; $\tau_{\pi - \psi_0}^\ell(r, \Omega)$ will be determined by using existing global models of the gravitational potential.

8.3 A continuous downward continuation problem

Using the Poisson integral (8.17), the downward continuation problem (8.8)–(8.10) means to find function $\tau^\ell(R, \Omega)$ satisfying Fredholm's integral equation of the 1st kind,

$$\boxed{\frac{1}{4\pi} \int_{\Omega_0} \tau^\ell(R, \Omega') K^\ell(R + H(\Omega), \psi, R) d\Omega' = f(\Omega) ,} \tag{8.22}$$

where known function $H(\Omega) > 0$ is the height of the Earth's surface above the geoid reckoned along the geocentric radius. (Note that once $H(\Omega) = 0$ in a particular direction Ω, the Poisson kernel becomes the Dirac delta function, and eqn.(8.22) has a simple solution $\tau^\ell(R, \Omega) = f(\Omega)$.)

The solution of integral eqn.(8.22) can be expressed in terms of the complete normalized system of eigenfunctions $\{u_i(\Omega)\}$, $\{v_i(\Omega')\}$, and the eigenvalues $\{\lambda_i\}$ of Poisson's kernel. The eigenvalue expansion of $K^\ell(R + H(\Omega), \psi, R)$ can be written as the following infinite sum

$$K^\ell(R + H(\Omega), \psi, R) = \sum_{i=1}^{\infty} u_i(\Omega) \lambda_i v_i^*(\Omega'), \tag{8.23}$$

where asterisk denotes complex conjugation. The solution of eqn.(8.22) then reads

$$\tau^\ell(R,\Omega') = \sum_{i=1}^{\infty} \frac{f_i}{\lambda_i} v_i(\Omega') , \qquad (8.24)$$

where f_i are the Fourier coefficients of the expansion of $f(\Omega)$ by means of $u_i(\Omega)$, i.e.,

$$f(\Omega) = \sum_{i=1}^{\infty} f_i u_i(\Omega) . \qquad (8.25)$$

Numbering λ_i such that they create a non-increasing series, $\lambda_1 \geq \lambda_2 \geq ...,$ eigenvalues λ_i of Poisson's kernel are non-negative and approach zero when $i \to \infty$,

$$\lambda_i \geq 0 , \qquad\qquad \lim_{i\to\infty} \lambda_i = 0 . \qquad (8.26)$$

For instance, if $H(\Omega) = H_0 = const. > 0$ all over the full solid angle Ω_0, then

$$\lambda_i = \left(\frac{R}{R+H_0} \right)^{i+1} , \qquad (8.27)$$

and $u_i(\Omega)$, $v_i(\Omega')$ are complex fully normalized spherical harmonics.

The series on the right-hand side of eqn.(8.24) converges (and, thus, it represents the solution of eqn.(8.22)), if and only if the Picard condition holds (Groetsch, 1984, sect.1.2; Kondo, 1991, sect.6.9; Hansen, 1992):

$$\boxed{\sum_{i=1}^{\infty} \left(\frac{f_i}{\lambda_i} \right)^2 < \infty .} \qquad (8.28)$$

The Picard condition says that starting from some point in the summation in eqn.(8.28), the absolute value of Fourier coefficients f_i must decay faster than the corresponding eigenvalues λ_i.

8.4   Discretization

So far, the downward continuation problem has been formulated in a continuous way. This means that an infinite number of coefficients f_i should be determined from boundary functional $f(\Omega)$ given continuously at all points on the Earth's surface. But, in practice, the boundary functionals $f(\Omega)$ is measured at discrete points only, and hence, the amount of coefficients f_i which can be determined from such discrete data is finite. The infinite series (8.25) must be truncated at some finite cut-off degree. Such a truncation represents a certain type of regularization because a high frequency part of $\tau^\ell(R,\Omega)$, determination of which makes the problem unstable, is excluded from the solution. A question arises whether the regularization of the solution by truncation its high-frequency part is efficient enough to stabilize the solution of Fredholm's integral equation (8.22).

Or, more specifically, up to which cut-off degree the solution of eqn.(8.22) is still numerically stable. Since truncation in a spectral domain corresponds to a given resolution in a spatial domain, we may also look for the smallest spatial grid step $\Delta\Omega$ for which a discretized form of eqn.(8.22) has still a stable solution.

To answer the above question, let us rewrite the integral equation (8.22) in a discrete form and find a discrete solution. Let observations of boundary functional $f(\Omega)$ run over a regular angular grid with grid step $\Delta\Omega = (\Delta\vartheta, \Delta\lambda)$, where $\Delta\vartheta$ and $\Delta\lambda$ are grid steps in latitude and longitude, respectively. Let observations result in a finite set of discrete values $f_i = f(\Omega_i)$, $i = 1, ..., N$. The solution $\tau^\ell(R, \Omega)$ may then be parameterized by discrete values $\tau^\ell(R_i, \Omega_i)$, $i = 1, ..., N$, evaluated over the same angular grid as observations f_i. Poisson's integral on the left-hand side of eqn.(8.22) can be computed using a numerical quadrature with the nodes coinciding with grid intersections and weights w_i.

Moreover, in order to transform the integral equation (8.22) into a system of linear algebraic equations, we will use the decomposition of Poisson's integral into three constituents, see eqn.(8.21). The smallest one, far-zone contribution $\tau^\ell_{\pi-\psi_0}(r, \Omega)$, is assumed to be computed in advance, before solving a discrete problem, and hence, $\tau^\ell_{\pi-\psi_0}(r_i, \Omega_i)$ will appear on the right-hand sides of eqns.(8.22). Such handling with far-zone contribution is enabled by its small size compared to near-zone contributions $\tau^\ell_0(r, \Omega)$ and $\tau^\ell_{\psi_0}(r, \Omega)$. An outline how to compute $\tau^\ell_{\pi-\psi_0}(r_i, \Omega_i)$ employing existing global models of the gravitational field is introduced in Appendix B.8 after eqn.(B.8.5).

The discrete form of Fredholm's integral equation (8.22) of the 1st kind then reads

$$\sum_{j=1}^{N} A_{ij}\tau^\ell(R, \Omega_j) = f(\Omega_i) - \tau^\ell_{\pi-\psi_0}(r_i, \Omega_i) , \qquad i = 1, ..., N , \qquad (8.29)$$

where the diagonal elements of square $N \times N$ matrix A_{ij} are equal to

$$A_{ii} := d^\ell(r_i, \psi_0, R) - \frac{1}{4\pi} \sum_{\substack{j=1 \\ j \neq i}}^{N} w_j K^\ell(r_i, \psi_{ij}, R) , \qquad (8.30)$$

and $d^\ell(r_i, \psi_0, R)$ is given by eqn.(B.8.17) taken at the i-th grid point. The off-diagonal elements A_{ij}, $i \neq j$, read

$$A_{ij} := \begin{cases} \dfrac{1}{4\pi} w_j K^\ell(r_i, \psi_{ij}, R) , & \text{if } \psi_{ij} \leq \psi_0, \\[2ex] 0 & \text{otherwise} . \end{cases} \qquad (8.31)$$

Furthermore, we have abbreviated notations introducing $r_i = R + H(\Omega_i)$.

The Picard condition (8.28) for a discrete downward continuation problem becomes

$$\sum_{i=1}^{N} \left(\frac{f_i}{\lambda_i}\right)^2 < \infty , \qquad (8.32)$$

where λ_i are the eigenvalues of matrix A, and f_i are the Fourier coefficients of the expansion of $f(\Omega)$ with respect to the eigenvector of matrix A (Do not change f_i with $f(\Omega_i)$). From a purely mathematical point, the discrete Picard condition (8.32) is always satisfied and we need not bother with the stability of the solution. However, discrete problem suffers from a combination of measurement errors, discretization errors, and roundoff errors, and the solution to the discrete problem may become extremely sensitive to these errors. Hence, in practice, before solving the discrete downward continuation problem, it should be checked whether the Fourier coefficients f_i decay faster in average than the eigenvalues λ_i for high frequencies i.

8.5   Jacobi's iterations

In our initial investigations (Vaníček et al., 1996), we solved the system of linear algebraic equations (8.29) iteratively setting up and storing matrix A preliminary on a computer hard disc. This led to operate with a huge matrix of system, and consequently, a huge memory of a computer was required to solve eqns.(8.29). However, this obstacle may be simple overcome employing Jacobi's (or Ritz's) iteration approach for solving a system of linear algebraic equations (Ralston, 1965, sect.9.7-1; Rektorys, 1968, sect.30.2). The basic idea behind Jacobi's iteration solution is quite simple. A brief description is as follows. Let matrix notation of system of eqns.(8.29) be written as

$$Ax = y , \tag{8.33}$$

where y is a known vector composed from the right-hand sides of eqns.(8.29), and x consists of unknowns $\tau^\ell(R, \Omega_j)$. Let matrix A be arranged to the form

$$A = I - B , \tag{8.34}$$

where I is the unit matrix of order N. Substituting eqn.(8.34) into (8.33), we arrive at

$$x = y + Bx . \tag{8.35}$$

The system of eqns.(8.35) may be solved iteratively starting with

$$x \equiv x_0 = y . \tag{8.36}$$

At the k-th stage of iteration $(k > 0)$ we are carrying out x_k according to equation

$$x_k = Bx_{k-1} . \tag{8.37}$$

When $|x_k - x_{k-1}|$ is less than some tolerance ϵ, we can stop iterating. The result of this operation yields the solution of eqn.(8.33):

$$x = y + \sum_{k=1}^{K} x_k , \tag{8.38}$$

where K is number of iteration steps.

In the k-th stage of iteration, we have to compute the product $\boldsymbol{B}\boldsymbol{x}_{k-1}$ which is nothing else but a discretized form of Poisson's integral (8.17) (with a slightly modified kernel for the diagonal elements) applied to $\boldsymbol{x}_{k-1}$. Thus, employing Jacobi's iterative approach there is no necessity to evaluate and store matrix $\boldsymbol{A}$ or $\boldsymbol{B}$ separately, but only Poisson's integral is to be carried out K-times by a method of numerical quadrature. Note that the above Jacobi iterative scheme is a discretized version of the iterative method proposed by Heiskanen and Moritz (1967, sect.8-10).

8.6   Numerical tests

8.6.1   Analysis of conditionality

The first problem we will investigate numerically concerns the conditionality of the matrix $\boldsymbol{A}$. Judging from the instable behaviour of the solution to the continuous downward continuation problem, the matrix $\boldsymbol{A}$ may become fairly ill-conditioned or even singular. In such a case, the solution to the discrete problem would be unstable, and other type of regularization, more powerful than simple cutting the solution at the Nyquist frequency $\pi/\Delta\Omega$, $\Delta\Omega$ being a grid step size, would have to be applied to.

To analyse the conditionality of matrix $\boldsymbol{A}$, we will use the eigenvalue decomposition technique. According to this method, matrix $\boldsymbol{A}$ can be decomposed to product of three matrices

$$\boldsymbol{A} = \boldsymbol{U}\boldsymbol{\Lambda}\boldsymbol{V}^T \, , \qquad (8.39)$$

where matrices $\boldsymbol{U}$ and $\boldsymbol{V}$ are column-orthogonal (matrix $\boldsymbol{V}$ is also row-orthogonal), and the diagonal matrix $\boldsymbol{\Lambda}$ consists of eigenvalues λ_i, $i = 1, ..., N$. If any eigenvalue is zero or very small, then the matrix $\boldsymbol{A}$ is singular or near to singular. As usual, we will measure the conditionality of a matrix by the condition number κ, defined as the ratio of the largest of the λ_i's to the smallest of the λ_i's (Wilkinson, 1965, sect.2.30):

$$\kappa = |\lambda_{max}|/|\lambda_{min}| \, . \qquad (8.40)$$

A matrix is singular if its condition number is infinite. To make decision whether a matrix is ill-conditioned, we should compare the reciprocal value of the condition number κ with the machine's floating point precision ϵ (for example, $\epsilon \doteq 10^{-6}$ for single precision), and say that problem is ill-conditioned if $1/\kappa$ multiplied by a constant is comparable or less than machine precision ϵ. For the constant one may choose a larger dimension of the matrix, or the square root of this number, or another constant; that starts getting into hardware-dependent question.

The conditionality of our matrix $\boldsymbol{A}$, eqns.(8.30)–(8.31), was studied over an area $3° \times 6°$, delimited by latitudes $47°$N and $50°$N, and by longitudes $242°$E and $248°$E. This area covers a particularly rugged part of the Canadian Rocky

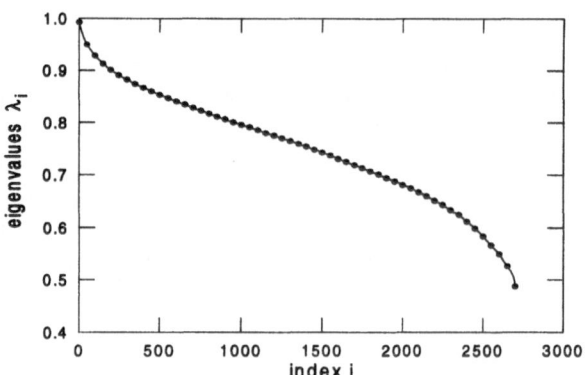

Figure 8.1: Eigenvalue spectrum of matrix $\boldsymbol{A}$ for the region of Columbia Mountains ($47° \leq \phi \leq 50°$, $242° \leq \lambda \leq 248°$) with the $5' \times 5'$ grid of topographical heights.

Mountains, the chain of Columbia Mountains. The mean $5' \times 5'$ topographical heights range between 503 and 2425 metres.

Figure 8.1 is a plot of eigenvalues λ_i of matrix $\boldsymbol{A}$ ordered according to their size. We can observe that the values decrease smoothly with the condition number $\kappa = 2.05$. This behaviour contrasts to that expected from the stability analysis of the continuous case, namely, matrix $\boldsymbol{A}$ is well-conditioned. Unfortunately, the computation of eigenvalues could not be carried out for a larger area because of huge consuming computational time and memory; the above example consumed nearly 45 hours of total CPU time on an HP-715 workstation with 64 MB of the internal memory.

We shall thus attempt to estimate the extremal eigenvalues. In accordance with Gerschgorin's theorem (Wilkinson 1965, sect.2.13), it holds

$$\lambda_{max} \leq \max_i \sum_{j=1}^{N} |A_{ij}| \,, \tag{8.41}$$

and

$$\lambda_{min} \geq \min_i (A_{ii} - \sum_{\substack{j=1 \\ j \neq i}}^{N} |A_{ij}|) \,. \tag{8.42}$$

These estimates can readily be evaluated in the case that Poisson's integration is taken over the full solid angle Ω_0, and the integration kernel is the full Poisson kernel $K(r, \psi, R)$. In such a case the matrix A_{ij} takes the form

$$A_{ii} = \frac{R}{r_i} - \frac{1}{4\pi} \sum_{\substack{j=1 \\ j \neq i}}^{N} w_j K(r_i, \psi_{ij}, R) \,, \tag{8.43}$$

$$A_{ij} = \frac{1}{4\pi} w_j K(r_i, \psi_{ij}, R) \,, \qquad \text{if } i \neq j \,. \tag{8.44}$$

From these equations immediately follows that

$$\sum_{j=1}^{N} |A_{ij}| = \frac{R}{r_i} , \tag{8.45}$$

and, by eqn.(8.41), the upper limit of the eigenvalues of matrix $\boldsymbol{A}$ is

$$\boxed{\lambda_{max} < 1 ,} \tag{8.46}$$

since $r_i > R$ for any i. The estimate (8.46) is in the full agreement with the maximum size of eigenvalues of matrix $\boldsymbol{A}$ plotted in Figure 8.1 because the influence of removing long-wavelength harmonics $j = 0, ..., \ell-1$, from Poisson's kernel $K(r, \psi, R)$ (in order to create the spheroidal Poisson kernel $K^{\ell}(r, \psi, R)$) on the largest eigenvalue λ_{max} is tiny. Similarly, the far-zone contribution to Poisson's integral has a very small impact on the property of matrix $\boldsymbol{A}$ (see the discussion after eqn.(8.21)), and therefore, we can include it into estimates of the extremal eigenvalues of $\boldsymbol{A}$.

To get the lower limit of the eigenvalues λ_i according to eqn.(8.42), we can proceed as follows:

$$A_{ii} - \sum_{\substack{j=1 \\ j \neq i}}^{N} |A_{ij}| = \frac{R}{r_i} - \frac{1}{2\pi} \sum_{\substack{j=1 \\ j \neq i}}^{N} w_j K(r_i, \psi_{ij}, R) = \frac{R}{r_i} - \frac{1}{2\pi} \int_{\Omega_0 \setminus \Omega_{\psi_1}} K(r_i, \psi, R) d\Omega , \tag{8.47}$$

where spherical cap Ω_{ψ_1} surrounds the computation point and has the radius which is equal to the minimum of grid step sizes $\Delta\vartheta$ and $\Delta\lambda$. With the help of eqn.(B.8.11), we further have

$$\frac{1}{4\pi} \int_{\Omega_0 \setminus \Omega_{\psi_1}} K(r_i, \psi, R) d\Omega = \frac{R}{r_i} - \frac{r_i + R}{2r_i}\left(1 - \frac{r_i - R}{\ell_i(\psi_1)}\right) , \tag{8.48}$$

where distance $\ell_i(\psi_1)$, given by eqn.(B.8.12), now becomes:

$$\ell_i(\psi_1) := \sqrt{r_i^2 + R^2 - 2r_i R \cos\psi_1} . \tag{8.49}$$

Substituting eqn.(8.48) into (8.47), we get

$$A_{ii} - \sum_{\substack{j=1 \\ j \neq i}}^{N} |A_{ij}| = 1 - \frac{r_i + R}{r_i}\frac{r_i - R}{\ell_i(\psi_1)} . \tag{8.50}$$

Realizing that difference $r_i - R$ is height H_i of the i-th surface point above the geoid (reckoned along the geocentric radius), and putting approximately $(r_i + R)/r_i \doteq 2$, we get

$$A_{ii} - \sum_{\substack{j=1 \\ j \neq i}}^{N} |A_{ij}| = 1 - 2\frac{H_i}{\ell_i(\psi_1)} . \tag{8.51}$$

In practical applications, ψ_1 is always a small angle. Therefore, we can apply the planar approximation (see Appendix A.8) to the last term, and write

$$\frac{H_i}{\ell_i(\psi_1)} \approx \frac{H_i}{\sqrt{\ell_0^2(\psi_1) + H_i^2}} = \sin \beta_i \; , \tag{8.52}$$

where

$$\ell_0(\psi_1) := 2R \sin \frac{\psi_1}{2} \; , \tag{8.53}$$

and β_i is the angle between the horizont and the line connecting the i-th point on the Earth's surface with its neighbouring point on the geoid. (Note that β_i is not the angle of inclination of the terrain.) Eqn.(8.51) can now be arranged to the form

$$A_{ii} - \sum_{\substack{j=1 \\ j \neq i}}^{N} |A_{ij}| = 1 - 2 \sin \beta_i \; . \tag{8.54}$$

To get the lower limit of eigenvalues of matrix A, we have to take the minimum of the expression standing on the right-hand side of eqn.(8.54), i.e.,

$$\lambda_{min} \geq \min_i (1 - 2 \sin \beta_i) \; , \tag{8.55}$$

or equivalently,

$$\boxed{\lambda_{min} \geq 1 - 2 \max_i (\sin \beta_i) \; .} \tag{8.56}$$

The last inequality is very helpful for estimating the condition number κ of matrix A in the case when $\max_i(\sin \beta_i) < 1/2$, i.e., when $\max_i(\beta_i) < 30°$. Then, inequality (8.56) estimates the minimum eigenvalue by a positive number, and we can see whether λ_{min} approaches zero or not, i.e., whether the condition number κ significantly grows or not. Unfortunately, the criterion (8.56) is of little use once $\beta_i > 30°$. Then, λ_{min} is estimated by a negative number, and we cannot decide whether λ_{min} is close to zero or not.

Applying estimate (8.56) to our example shown in Figure 8.1, we have

$$\sin \beta_i \leq \frac{2.425}{\sqrt{(2 \times 6371 \times \sin 2.5' \times \cos 50°)^2 + (2.425)^2}} \doteq 0.377 \; , \tag{8.57}$$

where 2.425 is the maximum topographical heights (in km) in this region, 6371 is the mean radius of the Earth (in km), 2.5' is a half of the discretization step, 50° is the latitude of the northest terrain profile. By means of (8.56), we have

$$\lambda_{min} > 0.245 \; . \tag{8.58}$$

This estimate is in a full agreement with the minimum size of eigenvalues of matrix A plotted in Figure 8.1 since $\lambda_{min} \doteq 0.48$.

Next, we now attempt to provide another estimate of the condition number by analysing the stability of the solution to the discrete downward continuation problem for a model with constant topographical heights, $H(\Omega) = H_0 = const.$ all over the world (see sect.8.3). Assuming that data functional $f(\Omega)$ is only given in discrete points of a regular angular grid with grid step size $\Delta\Omega$, its spectral alias-free series is finite, truncated at degree j_{max}

$$f(\Omega) = \sum_{j=0}^{j_{max}} \sum_{m=-j}^{j} f_{jm} Y_{jm}(\Omega) \, , \qquad (8.59)$$

where f_{jm} are expansion coefficients, and $j_{max} < \pi/\Delta\Omega$, where $\pi/\Delta\Omega$ being the Nyquist frequency (e.g., Colombo, 1981). The solution to the discrete downward continuation problem for our simple model of a constant height now reads (see eqn.(8.24)),

$$\tau^{\ell}(R,\Omega) = \sum_{j=\ell}^{j_{max}} \sum_{m=-j}^{j} \left(\frac{R+H_0}{R}\right)^{j+1} f_{jm} Y_{jm}(\Omega) \, . \qquad (8.60)$$

Hence, $\tau^{\ell}(R,\Omega)$ becomes unstable once

$$\left(\frac{R+H_0}{R}\right)^{j_{max}+1} \approx \frac{1}{\epsilon} \, , \qquad (8.61)$$

or equivalently,

$$\left(\frac{R+H_0}{R}\right)^{\pi/\Delta\Omega} \approx \frac{1}{\epsilon} \, , \qquad (8.62)$$

where ϵ is a machine floating point precision or a constant chosen according to the remark after eqn.(8.40). Thus, the condition number κ in this particular case can simply be estimated as

$$\kappa \approx \left(\frac{R+H_0}{R}\right)^{\pi/\Delta\Omega} \, . \qquad (8.63)$$

In the case when H is not constant over the region under study, we can replace H_0 by the maximum topographical heights H_{max}. Then, such an estimate obviously overestimates the actual condition number, i.e., it is too pessimistic, and hence it holds

$$\boxed{\kappa \leq \left(\frac{R+H_{max}}{R}\right)^{\pi/\Delta\Omega}} \, . \qquad (8.64)$$

For the example in Figure 8.1, we have

$$\kappa \leq \left(\frac{6371+2.425}{6371}\right)^{180\times12} \doteq 2.28 \, . \qquad (8.65)$$

We have already learnt that the actual condition number is $\kappa = 2.05$, so that criterion (8.64) estimates κ fairly well. Let us consider another example and put the machine floating point precision $\epsilon = 10^{-6}$ (a single precision), and height $H_0 = 6$ km. Then estimate (8.64) tells that the problem is unstable as soon as a discretization step is smaller than $\Delta\Omega \doteq 50$ arcsec.

Summing up, the numerical example shown in Figure 8.1 demonstrates that cutting the solution to the downward continuation problem for the gravity field at the frequency prescribed a priori by the discretization step of gravity data and/or topographical heights is a very powerful tool for regularization of the solution. The discrete downward continuation problem may be well-posed even for a very rough terrain such as the Rocky Mountains. Nevertheless, the posedness of the discrete downward continuation problem should be treated separately for each specific case; making a grid of topographical heights denser and denser, there is a limit of a grid step size expressed by criterion (8.56) or (8.64) on the conditionality of the matrix A which breaks down the stable behaviour of the discrete downward continuation problem and the problem becomes ill-posed. Then some kind of regularization technique outlined in Introduction must be applied.

8.6.2   Analysis of convergency

Now, let us have a look at the convergency of Jacobi's iterations (8.33)-(8.38). The necessary and sufficient condition ensuring that Jacobi's iterative method converges is that the maximum eigenvalue of matrix B, $\lambda_{max}(B)$, is less than 1 (Ralston, 1965, sect.9.7-1),

$$\lambda_{max}(B) < 1 . \qquad (8.66)$$

The largest eigenvalue of matrix B can, for instance, be estimated by the Gerschgorin inequality (8.41). This estimate may be pessimistic yielding $\lambda_{max}(B)$ close to 1. That is why, we shall determine $\lambda_{max}(B)$ precisely by an iterative process called the power method (Ralston, 1965, sect.10.2). The idea of this method is simple. Choose a vector v_0 such that it has a non-zero component in the direction of the eigenvector associated to the maximum eigenvalue $\lambda_{max}(B)$ (if we happen to choose vector v_0 perpendicular to the eigenvector associated to the maximum eigenvalue $\lambda_{max}(B)$ and the approach does not work, we repeat it starting with a different v_0), and generate a set of vectors v_k according to prescription

$$c_k v_k = B v_{k-1} , \qquad (8.67)$$

where c_k is equal to the component of vector Bv_{k-1} of the largest size. Numbers c_k then converge to $\lambda_{max}(B)$,

$$\lim_{k \to \infty} c_k = \lambda_{max}(B) . \qquad (8.68)$$

In the k-th stage of iterations, we only need to compute vector Bv_k and not matrix B separately. Again, as in the case of Jacobi's iterative approach, this leads to carry out a discretized Poisson integral (8.17) by a method of numerical quadrature.

The maximum eigenvalue of matrix B has been determined for two areas. In <u>area A</u> delimited by latitudes 40°N and 76° N, and by longitudes 214°E and 258°E which covers the whole region of the Canadian Rocky Mountains,

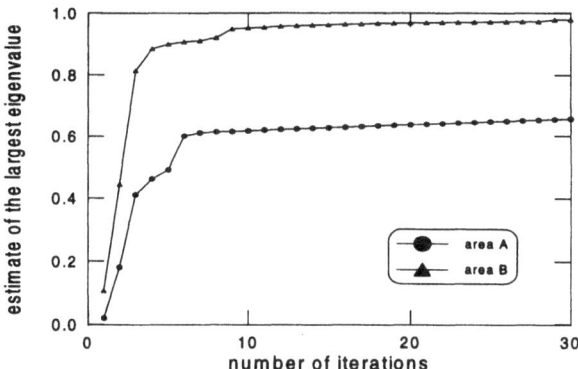

Figure 8.2: Sequence of numbers c_k, $k = 1, 30$, for our two areas. Figure shows that the limit of c_k for $k \to \infty$ approximately equals to 0.67 for area A, and 0.98 for area B.

the topographical heights are sampled as means in 5' by 5' grid. The height ranges between 0 and 3993 metres. The topographical heights in <u>area B</u> delimited by latitudes 47°N and 57°N, and longitudes 238°E and 248°, are grided much denser (30" by 60") than in area A. The height ranges between 0 and 3573 metres. This area covers a particularly rugged part of the Canadian Rocky Mountains, the chain of Columbia Mountains.

Figure 8.2 shows the result of iterations (8.67), i.e., the sequence of numbers c_k, $k = 1, ...30$. The limit of c_k's yielding $\lambda_{max}(B)$ is approximately equal to 0.67 for area A, and 0.98 for area B. The Jacobi iterative scheme will undoubtedly converge in the case of area A, while for area B with a very dense grid of topographical heights Jacobi's iterations will converge very slowly since the maximum eigenvalue is very close to 1. Let us have a look at the condition number κ of matrix A for area B. Criterion (8.56) yields

$$\lambda_{min} > 1 - 2 \, \frac{3.573}{\sqrt{(2 \times 6371 \times \sin 30'' \times \cos 57°)^2 + (3.573)^2}} \doteq -0.925 \, , \quad (8.69)$$

where 3.573 is the maximum topographical heights (in km) in this region, 6371 is the mean radius of the Earth (in km), 30" is a half of the discretization step in longitude direction, 57° is the latitude of the northest terrain profile. This is the case when the criterion (8.56) cannot help us to estimate the condition number κ since we cannot decide how close λ_{min} is to zero. Let us attempt to use criterion (8.64); for area B it reads

$$\kappa \leq \left(\frac{6371 + 3.573}{6371} \right)^{180 \times 120} \doteq 1.9 \times 10^5 \, . \quad (8.70)$$

(Note that this criterion gives $\kappa \leq 3.87(!)$ for area A.) Since 1.9×10^5 begins to approach the reciprocal value of machine floating point precision (for arithmetic operations in single precision), the downward continuation problem for area B is ill-posed. Searching the solution to the discrete downward continuation problem for area B by Jacobi's iterative scheme, we can run into serious difficulties connected with accumulating roundoff errors (Ralston, 1965, sect.9.7-3).

8.6.3   Power spectrum analysis of gravity anomalies

For regional geoid computation, the downward continuation problem should be solved over as a large area as possible in order to minimize the margin effect of truncated Poisson's integration (Vaníček et al., 1996). The discretization of the problem then leads to a matrix of huge dimensions, and, unfortunately, present-day computers are not able to carry out the eigenvalue analysis. Thus, in present, we cannot check directly the validity of the Picard condition (8.32) for a discrete downward continuation problem in geoid computation.

Nevertheless, we can ask about the spectral property of function $f(\Omega)$. Eqns. (8.6) and (8.11) show that function $f(\Omega)$, and hence also its Fourier's coefficients f_i, depends on the way of compensation of topographical masses. For the purpose of geoid computation, we may, in principal, employ any compensation model the gravitational field of which is harmonic in the space outside the geoid. Thus, we may question which model of the compensation of topographical masses reduces a high-frequency part of $f(\Omega)$ most efficiently? Namely, such a model will be most convenient for solving the problem of downward continuation of gravity in geoid height computation because the discrete Picard condition (8.32) will be satisfied in the best possible way.

As a matter of fact, by a suitable choice of the mass density of a compensation model, we can, in principle, achieve that function $f(\Omega)$ is identically equal to zero (Moritz, 1990, sect.8.3),

$$f(\Omega) \equiv 0 \ . \tag{8.71}$$

The compensation density of such a model is looked for by the deconvolution technique leading to necessity to solve a system of linear algebraic equation for the parameters of compensation density. We again run into the same problem as above, namely, to solve a huge system of linear algebraic equations, particularly, when a discretization step of $f(\Omega)$ is tiny. We will thus not effort to construct an 'ideal' compensation density, ensuring that eqn.(8.71) holds, but we will study the influence of some idealized compensation models with a simply constructible compensation density on the spectral property of function $f(\Omega)$ and attempt to find a compensation model which reduces a high frequency part of $f(\Omega)$ in the most efficient way.

According to the way of the isostatic compensation or the condensation of topographical masses (see Chapter 3), function $-f(\Omega)/[R+H(\Omega)]$ will be called the Airy-Heiskanen gravity anomaly, the Pratt-Hayford gravity anomaly or Helmert's gravity anomaly, respectively. To analyse the spectral contents of respective gravity anomalies, we have calculated their power spectra by using the autoregressive spectral method, also known as the maximum entropy method (e.g., Marple, 1987, chapt.8). The spectral power estimates carried out by this method have a better frequency resolution and increased signal detectability compared to the classical power spectral estimation based on the periodogram. The 5' by 5' free-air gravity anomaly data and the 5' by 5' topographical heights (needed for computing the topographical effects) have been considered for the rugged region of the Canadian Rocky Mountains delimited by latitudes $\phi = 40°N$ and $76°N$, and longitudes $\lambda = 214°E$ and $258°$ (the area A from the preceding section).

In order to show the resulting power spectra as transparent as possible, we have plotted them along chosen longitudinal profiles rather than as isolines or surface plots. Figures 8.3–8.6 shows the input topographical and free-air gravity data, the Helmert and Airy-Heiskanen gravity anomalies, and corresponding power spectral densities (including the power spectral density of the Pratt-Hayford compensation model) estimated by the maximum entropy method along four longitudinal profiles crossing the Rocky Mountains. Inspecting these figures, and also the power spectra along other profiles (not shown here), we can make the following conclusions.

• In all the cases we have investigated, the Airy-Heiskanen compensation model reduces high-frequency components of the free-air gravity anomalies in the most efficient way.

• There are profiles (see, e.g., Figure 8.3 and 8.4), where the Helmert's 2nd condensation technique reduces a short-wavelength part of the free-air gravity anomalies only slightly, while the Airy-Heiskanen model removes this part of spectra nearly completely.

• We can find the profiles (e.g., Figure 8.5) along which the Pratt-Hayford model has a similar damping effect on high-frequencies of free-air gravity anomalies as the Airy-Heiskanen model, but there are regions represented, for instance, by Figure 8.6, where the Airy-Heiskanen model reduces short-wavelengths more efficiently than the Pratt-Hayford model.

8.6.4   Downward continuation of gravity anomalies

Let us finally solve the discretized Fredholm integral equation (8.29) for the respective surface gravity anomalies, and determine grid values of function $\tau(R, \Omega)$. We use a simple Jacobi's iterative scheme which enables to find the solution without storing a huge matrix of the system of equations. We have learnt that for a region of the Canadian Rocky Mountains with surface observables discretized in a $5' \times 5'$ grid, the matrix of linear system of equations is well-conditioned (the condition number is equal to 3.9). Moreover, the matrix associated to Jacobi's iterations is contractive with the largest eigenvalue equal to 0.67. These two facts imply that the Jacobi iteration process converges to the solution which is not contaminated by large roundoff errors. To reach the absolute accuracy of the result of about 0.1 mGal, the number of iterations does not exceed 30. The result

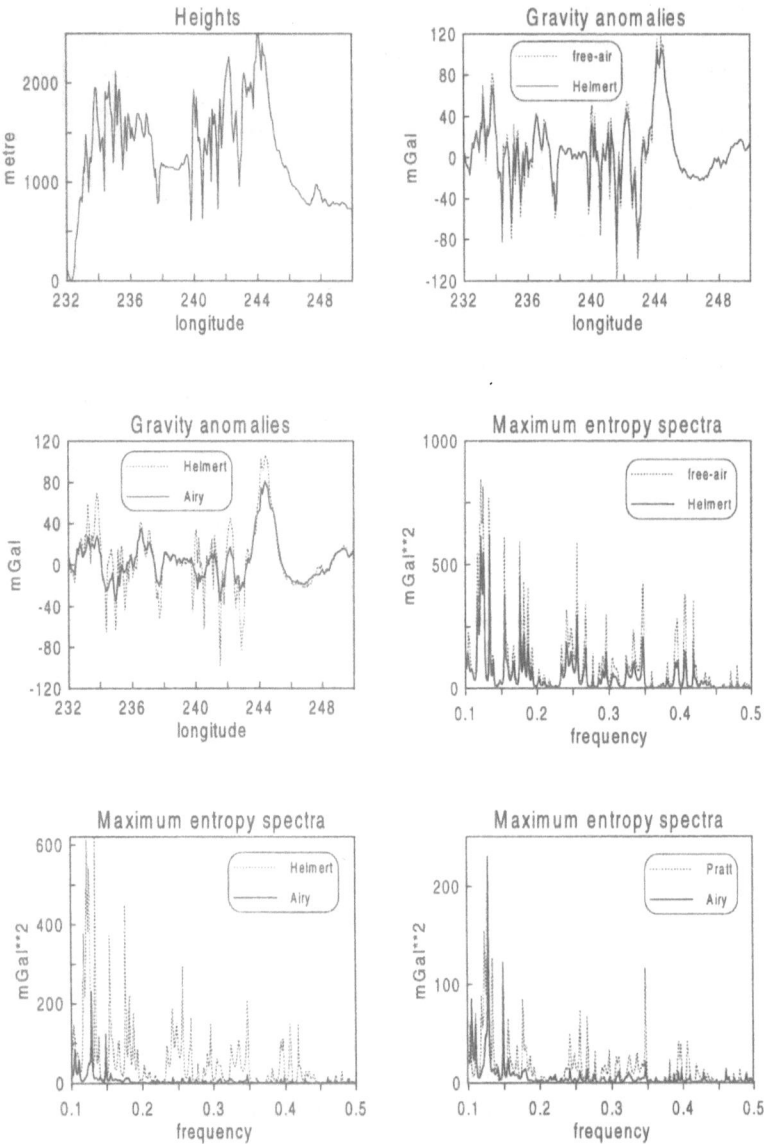

Figure 8.3: The topographical heights, the free-air gravity anomalies, the Helmert and Airy-Heiskanen gravity anomalies plotted along the longitudinal profile $\phi = 51.46°N$. The power spectral densities of the respective gravity anomalies (including those of the Pratt-Hayford gravity anomalies) are estimated by the maximum entropy method. The frequency times sampling interval ($= 5$ arcmin) is drawn on the x-axis of power spectrum figures. Shown is an expanded portion of the full Nyquist frequency interval (which would extend from zero to 0.5).

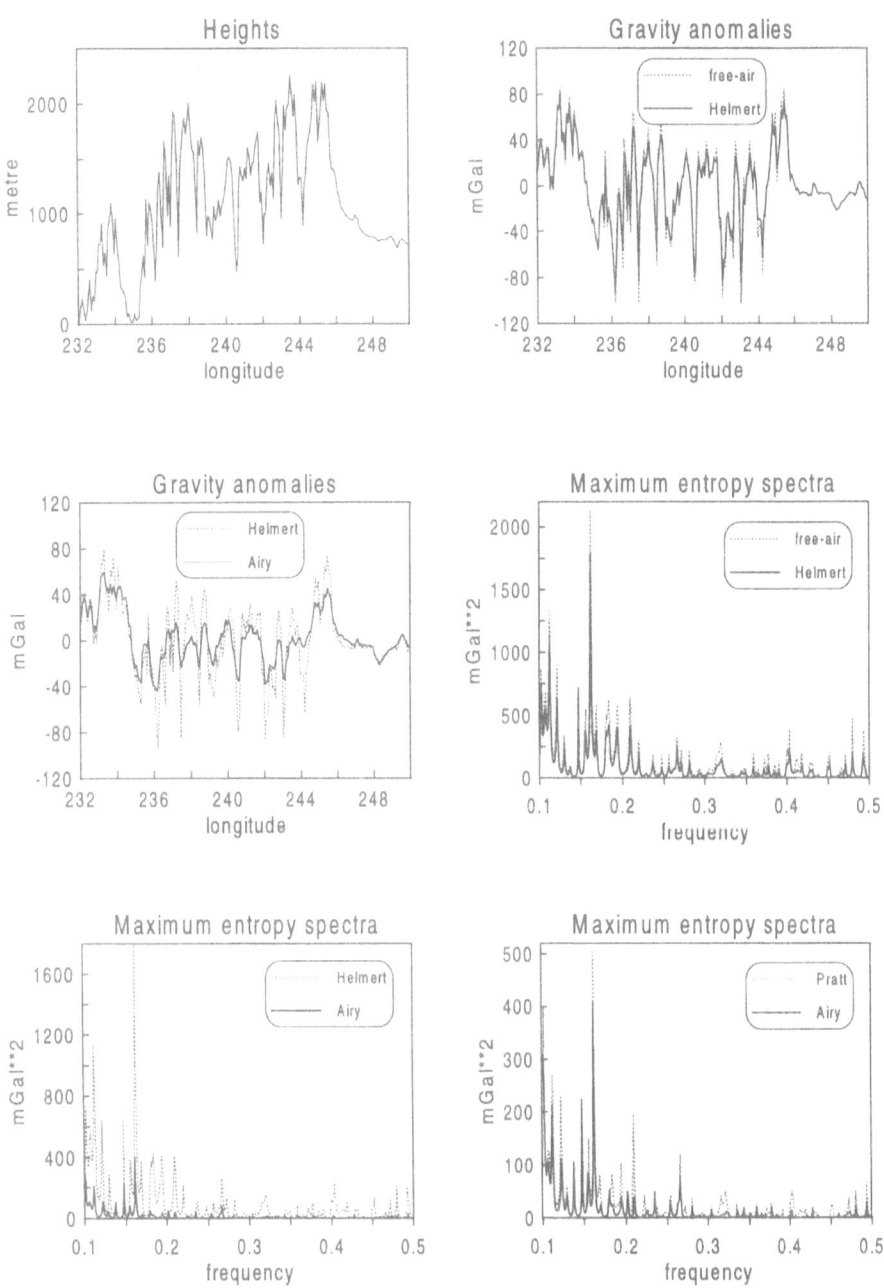

Figure 8.4: The same as Figure 8.3 for $\phi = 50.13°$N.

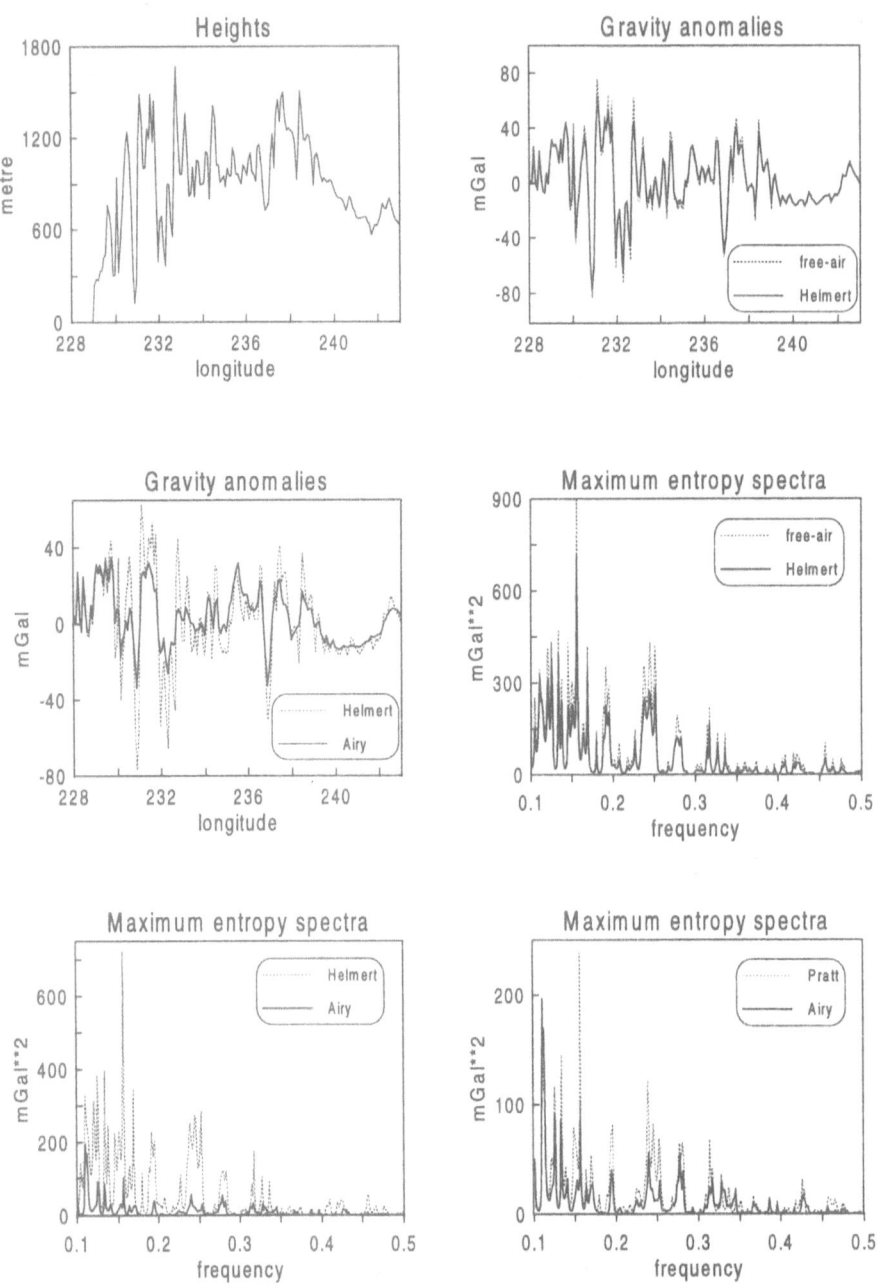

Figure 8.5: The same as Figure 8.3 for $\phi = 55.21°$N.

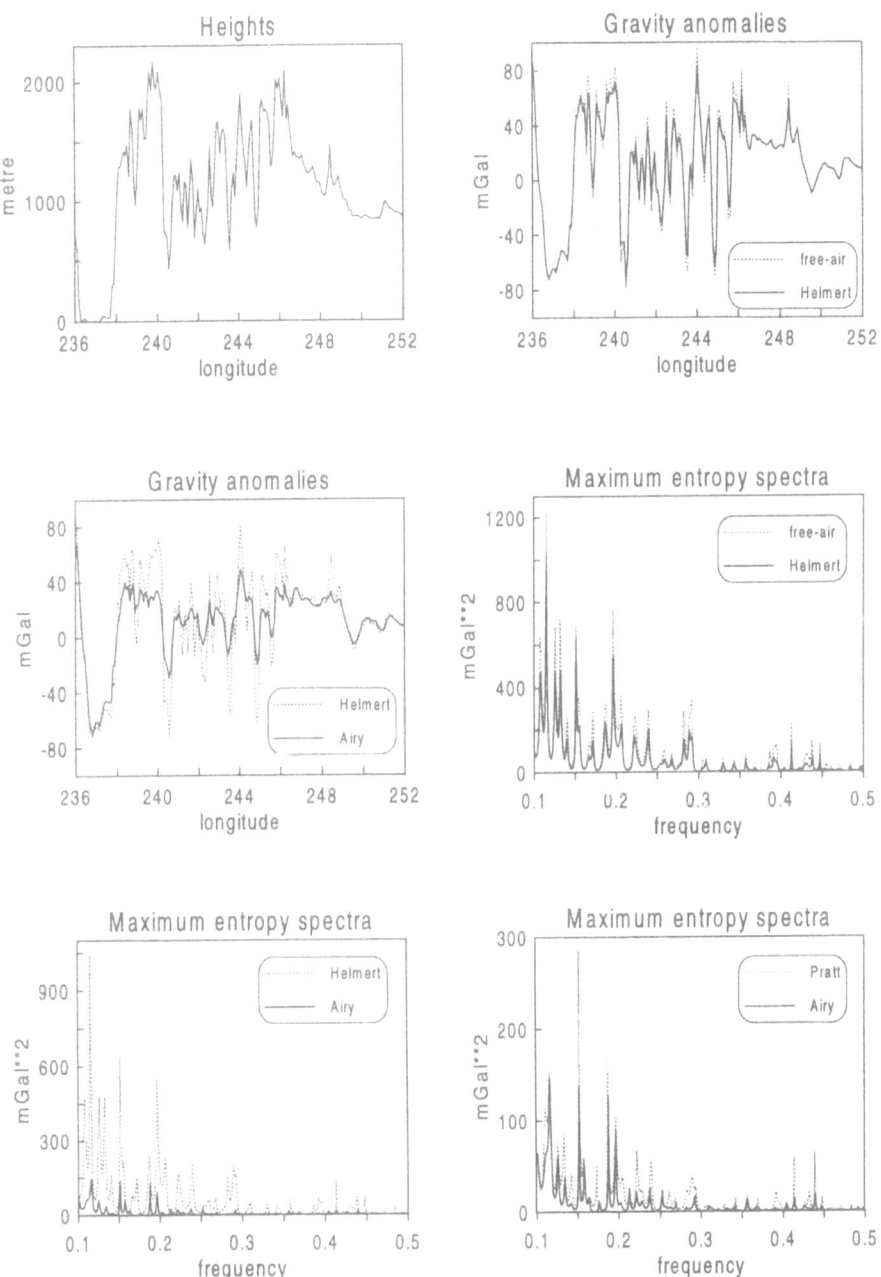

Figure 8.6: The same as Figure 8.3 for $\phi = 48.96°$N.

of this discrete downward continuation procedure will be presented by means of function $D\Delta g(\Omega)$,

$$D\Delta g(\Omega) := \frac{1}{R + H(\Omega)} \left[\tau(R, \Omega) - f(\Omega)\right] , \tag{8.72}$$

which can be understood to be a <u>downward continuation of gravity anomalies</u>.

Figure 8.7 plots the downward continuation of the Helmert and Airy-Heiskanen gravity anomalies along the four longitudinal profiles shown in Figures 8.3–8.6. We can observe that most of the power of $D\Delta g(\Omega)$ is of a very short wavelength. Moreover, as expected from the power spectral analysis, the amplitudes of $D\Delta g(\Omega)$ for the Airy-Heiskanen gravity anomalies are smaller (often significantly smaller) than the corresponding amplitudes of $D\Delta g(\Omega)$ for the Helmert gravity anomalies.

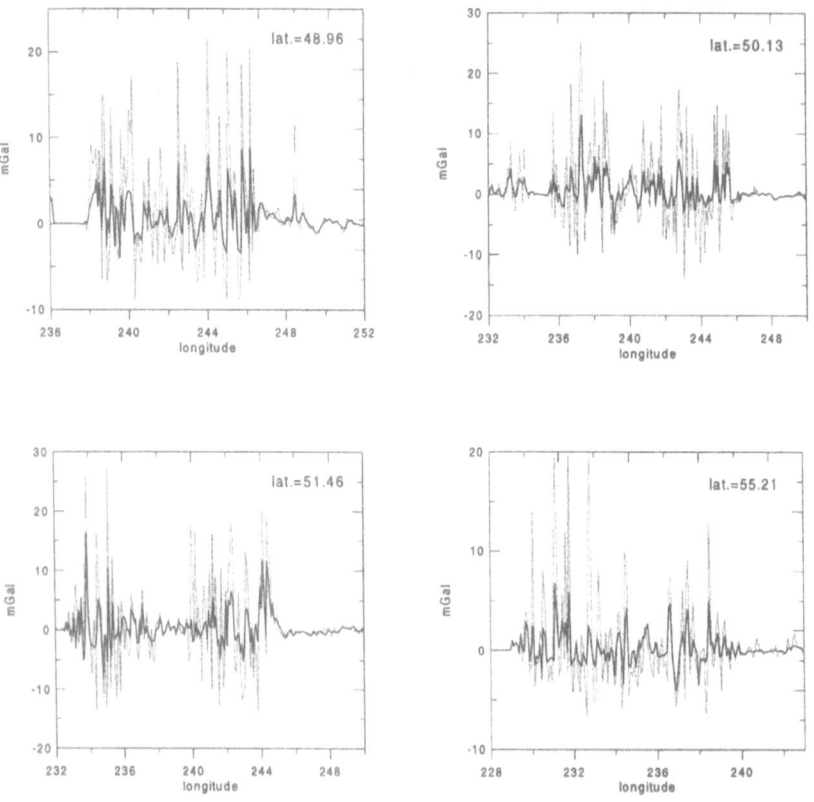

Figure 8.7: The downward continuation (in mGal) of the Helmert gravity anomalies (dotted lines) and the Airy-Heiskanen gravity anomalies (full lines) along four longitudinal profiles across the Canadian Rocky Mountains.

8.7   Conclusion

A crucial point of Stokes's method for geoid computation is to continue the gravity observations from the Earth's surface to the geoid. By solving this problem, high-frequency components of the surface gravity data are amplified which may make the problem unstable. The consequence is that errors contaminated data together with roundoff errors result in high frequency oscillations of the solution. This statement drawn for a continuous case is somehow weakened for a discrete downward continuation problem when surface observations are given in a discrete points only. Discretizing the solution with the same grid step as the observations then means that a high frequency part of the solution, determination of which makes the problem unstable, is excluded from the solution; the cut-off frequency is given by the Nyquist frequency corresponding to a spatial grid step size.

This chapter was motivated by a question on the smallest grid step size by which the discretized downward continuation problem is still well-posed. Or alternatively, we looked for the grid step size which surely breaks down the well-posedness of the problem. Note that there is a difference between these two questions since it is not possible to determine a sharp limit between the well-posedness and ill-posedness of a problem. To answer these questions, we have discretized the Poisson integral in a similar fashion as Bjerhammar (1987) or Vaníček at al. (1996), set up the system of linear algebraic equations, and studied the conditionality of the system matrix for a particularly rugged terrain of the Canadian Rocky Mountains.

The only reliable way to treat the conditionality of a matrix is to investigate its eigenvalues and evaluate the ratio of the largest to the smallest eigenvalues. We have followed this hint and employed the subroutine SVDCMP (Press et al., 1992, sect.2.6) to carry out the eigenvalues of the matrix of system of equations composed from discretized Poisson's kernel. Despite that this subroutine belongs to the most efficient techniques for finding all eigenvalues of a matrix, it has to operate with a matrix under study. This is a principal limitation of their usage for treating the stability of the discrete downward continuation problem. Namely, the downward continuation problem should be solved over as a large area as possible to minimize the margin effect of truncated Poisson's integration (Vaníček et al. 1996). Thus the eigenvalue software package is forced to handle with a matrix of a huge dimension, particularly, when a discretization step is tiny. On a HP-715 workstation with 64 MB of the memory, we have succeeded in carrying out the eigenvalue analysis of the system matrix for area $3° \times 6°$ with topographical $5' \times 5'$ heights (the dimensions of matrix to be analyse were 2592×2592). The result of the analysis plotted in Figure 8.1 is certainly a surprise; the condition number is 2.05 which means that the matrix of system of linear equations for the discrete downward continuation problem is fairly well-conditioned.

The impossibility to carry out the eigenvalue analysis for the problem over a larger area with a denser grid step of topographical heights made us propose simplified criteria (8.56) and (8.64) for making a decision whether the matrix

A of associated linear algebraic equations is well-conditioned or not. The price paid for their simplicity is that they yield more pessimistic estimates on the conditionality of this matrix than proper eigenvalue analysis. Nevertheless, these criteria provide a fairly reliable estimates on the condition number of matrix *A* for our test area A. Realizing that the transition between the stability and the instability of the problem is rather broad, the simplified criteria (8.56) or (8.64) may give a fairly good view into the stability property and help us to decide whether the discrete downward continuation problem for geoid determination is stable or not.

The second question we dealt with which is closely related to the posedness of the downward continuation problem concerns the convergency of Jacobi's iterative scheme suggested for searching the solution to the discrete downward continuation problem. Our analysis of the convergency has been based on the fact that the matrix *B* mediating the iterative solution has to be contractive which means that its largest eigenvalue must be smaller than 1. We have used the Gerschgorin estimate (8.41) of the maximum eigenvalue, but it has yielded a pessimistic estimate equal to 1. Therefore, the largest eigenvalue of matrix *B* has been looked for by the power method. As a result, we have found that the largest eigenvalue of matrix *B* is 0.67 for our test area A with topographical $5' \times 5'$ heights. So, we can conclude that Jacobi's iterations will surely converge in this case. Moreover, keeping in mind that the problem is well-posed, the result will not be contaminated by large roundoff errors.

On the other hand, the analysis of the largest eigenvalue of matrix *B* established for area B with discretization $30'' \times 60''$ shows that Jacobi's iterative scheme will converge very slowly. Using criterion (8.64) to test the conditionality of matrix *A*, we have found that the problem for area B is ill-posed. In such a case, searching the solution by Jacobi's iterative scheme may run into serious difficulties because of the accumulation of the roundoff errors.

Finally, we can answer the two questions raised at the beginning. The discrete downward continuation problem for geoid determination is undoubtedly well-posed and the solution may be looked for by Jacobi's iterative method once the grid step size of the surface observations as well as of the discrete solution is not smaller than 5 arcmin. This conclusion drawn for the Canadian Rocky Mountains will be valid anywhere else in the world, with perhaps the exception of Himalayas, since the Rockies represent one of the highest and roughest mountainous terrain pattern. On the contrary, discretizing the same region with grid step size $30'' \times 60''$, we have found that the downward continuation problem becomes unstable. In addition, Jacobi's iterations converge very slowly and cannot be, in fact, used for searching the solution.

One possible way how to regularize the solution consists in smoothing and dampening high frequency oscillations of the data by compensating the topographical masses in an appropriate way. For regional geoid computation, the present-day computers, are, unfortunately, not able to construct an 'ideal' compensation model which would remove completely a high-frequency part of surface gravity data. We have therefore studied the effect of highly idealized compensation and condensation models on the spectral contents of surface gravity data. We have found that for the region of the Canadian Rocky Mountains, the Airy-Heiskanen model reduces a high-frequency part of gravity data in the most efficient way. On the other hand, we have demonstrated that Helmert's 2nd condensation technique reduces high frequency oscillations in the worst way. We have found the areas ($\phi = 53° - 55°$N, $\lambda = 231° - 243°$E, and $\phi = 65° - 66°$N, $\lambda = 218° - 230°$E), where the Airy-Heiskanen method reduces high frequency components of surface gravity fairly significantly in the contrast to a very small reduction effect of Helmert's 2nd condensation technique. This is surely consequence of the fact that the latter technique is a pure mathematical tool for the description of compensation of topographical masses, whereas the Airy-Heiskanen model may, in some regions, approximate the actual compensation mechanism fairly well.

So, we recommend that the spectral analysis of respective gravity anomalies should be carried out before choosing a model of compensation of topographical masses for geoid height computation. The experience with the spectral analysis and the downward continuation procedure of the gravity observables from the Canadian Rocky Mountains indicates that rather the Airy-Heiskanen model than Helmert's 2nd condensation technique should be used to reduce high-frequency oscillations of surface gravity data.

A.8   Spherical radius of the near-zone integration cap

The radius ψ_0 of a near-zone spherical cap will be chosen such that the spatial distance $L(R + H, \psi, R)$, eqn.(A.5.10), outside the cap can be approximated by the distance ℓ_0, eqn.(3.57), with an error not larger than the error of planar approximation of distances (see sect.4.3). Taking the planar approximation of distance $L(R + H, \psi, R)$,

$$L(R + H, \psi, R) \approx \sqrt{\ell_0^2 + H^2} , \qquad (A.8.1)$$

and putting approximately $L(R + H, \psi, R) \approx \ell_0$ for $\psi > \psi_0$, we make an error of magnitude of H^2/ℓ_0^2. If we require that this error should not be larger than H/R, i.e.,

$$\frac{H^2}{\ell_0^2} \leq \frac{H}{R} , \qquad (A.8.2)$$

radius ψ_0 of the near-zone integration cap is given by the equation

$$\sin \frac{\psi_0}{2} = \frac{1}{2}\sqrt{\frac{H}{R}} . \qquad (A.8.3)$$

Since $H/R \ll 1$, we may put $\sin \psi/2 \doteq \psi/2$, and the condition (A.8.3) reduces to:

$$\psi_0 = \sqrt{\frac{H}{R}} \, . \tag{A.8.4}$$

The near-zone is then defined by those ψ's which are smaller than ψ_0. The smaller the height of the computation point, the smaller the near zone and the larger the far-zone. In the extreme case when the computational point is on Mount Everest, the near zone extends to the angular distance of about $\psi_0 = 2°$, and the far-zone from $2°$ to $180°$. When $H = 0$, there is no near-zone.

Under the condition (A.8.4), the Poisson kernel $K(r, \psi, R)$, eqn.(8.20), for integration points lying in the far-zone can be approximated as

$$K(r, \psi, R) \doteq \frac{2R^2 H}{\ell_0^3} \qquad \text{for } \psi > \psi_0 \, , \tag{A.8.5}$$

making an error which does not exceed the error of spherical approximation of the geoid.

B.8   Poisson's integration over near- and far-zones

As already introduced, for regional geoid determination it is advantageous to split the integration domain Ω_0 in the Poisson's integral (8.17) into near- and far-zone subdomains. The near-zone is created by a spherical cap surrounding the computation point while the rest of the full solid angle creates the far-zone subdomain. The radius ψ_0 of the spherical cap may be chosen by various manner; one possible choice is introduced in Appendix A.8 by eqn.(A.8.4).

Mathematically, splitting the integration domain Ω_0 as

$$\Omega_0 = \Omega_{\psi_0} \cup (\Omega_0 \setminus \Omega_{\psi_0}) \, , \tag{B.8.1}$$

where Ω_{ψ_0} is a spherical cap of radius ψ_0, the Poisson integral (8.17) reads

$$\tau^\ell(r, \Omega) = \bar{\tau}^\ell_{\psi_0}(r, \Omega) + \tau^\ell_{\pi - \psi_0}(r, \Omega) \, , \tag{B.8.2}$$

where term

$$\bar{\tau}^\ell_{\psi_0}(r, \Omega) := \frac{1}{4\pi} \int_{\Omega_{\psi_0}} \tau^\ell(R, \Omega') K^\ell(r, \psi, R) d\Omega' \tag{B.8.3}$$

expresses the contribution of the integration points lying in the near-zone spherical cap, and

$$\tau^\ell_{\pi - \psi_0}(r, \Omega) := \frac{1}{4\pi} \int_{\Omega_0 \setminus \Omega_{\psi_0}} \tau^\ell(R, \Omega') K^\ell(r, \psi, R) d\Omega' \tag{B.8.4}$$

expresses the contribution of the far-zone integration points. The crucial point consists in different ways of evaluating the particular contributions to Poisson's integral. The near-zone term $\overline{\tau}^{\ell}_{\psi_0}(r, \Omega)$ will be evaluated under the assumption that function $\tau^{\ell}(R, \Omega)$ is given continuously over spherical cap Ω_{ψ_0}; $\overline{\tau}^{\ell}_{\psi_0}(r, \Omega)$ will then be determined by computing integral (B.8.3) using some method of numerical integration. On the contrary, to determine the far-zone contribution $\overline{\tau}^{\ell}_{\pi-\psi_0}(r, \Omega)$, we will assume that function $\tau^{\ell}(R, \Omega)$ is approximated by a finite spherical harmonic series of the form,

$$\tau^{\ell}(R, \Omega) = \sum_{j=\ell}^{j_{max}} \sum_{m=-j}^{j} \tau^{\ell}_{jm} Y_{jm}(\Omega) , \tag{B.8.5}$$

where j_{max} is a finite cut-off degree; $\overline{\tau}^{\ell}_{\pi-\psi_0}(r, \Omega)$ will be determined by summation of a corresponding harmonic series. We thus assume that an estimate of spherical harmonics τ^{ℓ}_{jm} is available. Inspecting definition (8.7) of function τ^{ℓ}, we can observe that τ^{ℓ}_{jm}'s may be established by means of spherical harmonics of potential $T^{h,\ell}$. These harmonics can be set up by employing spherical harmonics of a global gravitational field and spherical harmonics of the Earth's topography (see Chapter 7).

B.8.1 Near-zone contribution

To begin with, let us evaluate the near-zone contribution $\overline{\tau}^{\ell}_{\psi_0}(r, \Omega)$. An increase in magnitude of the spheroidal Poisson kernel $K^{\ell}(r, \psi, R)$ when $\psi \to 0$ makes numerical computation of integral (B.8.3) difficult. There is a variety of numerical procedures how to treat this problem numerically (e.g., Shaofeng and Xurong, 1991; Gysen, 1994). The most straightforward way is to use the fact that

$$\int_{\Omega_{\psi_0}} K^{\ell}(r, \psi, R) d\Omega' < \infty \tag{B.8.6}$$

for $R \neq 0$ and $r \neq 0$, and to remove a small neighbourhood of the point $\psi = 0$ from the integration domain of integral (B.8.3); the separate contribution of this area is then evaluated analytically. Formally, eqn.(B.8.3) can be written as

$$\overline{\tau}^{\ell}_{\psi_0}(r, \Omega) = \frac{1}{4\pi} \tau^{\ell}(R, \Omega) \int_{\Omega_{\psi_0}} K^{\ell}(r, \psi, R) d\Omega' + \tau^{\ell}_{\psi_0}(r, \Omega) , \tag{B.8.7}$$

where

$$\tau^{\ell}_{\psi_0}(r, \Omega) := \frac{1}{4\pi} \int_{\Omega_{\psi_0}} \left[\tau^{\ell}(R, \Omega') - \tau^{\ell}(R, \Omega) \right] K^{\ell}(r, \psi, R) d\Omega' . \tag{B.8.8}$$

Let us investigate the limit for $\psi \to 0$ of the subintegral function in the angular integral (B.8.8). When $\psi \to 0$, then $\tau^{\ell}(R, \Omega') \to \tau^{\ell}(R, \Omega)$. It is reasonably to assume that function $\tau^{\ell}(R, \Omega)$ is bounded. This assumption means that there are no singularities of the gravitational field above the geoid. As the element $d\Omega'$ of the full solid angle in polar coordinates (ψ, α) is $d\Omega' = \sin\psi d\psi d\alpha$, the limit for $\psi \to 0$ of the subintegral function in eqn.(B.8.8) reads

$$\lim_{\psi \to 0}\left\{\left[\tau^\ell(R,\Omega') - \tau^\ell(R,\Omega)\right]K^\ell(r,\psi,R)\sin\psi\right\} =$$

$$= \tau^\ell(R,\Omega)\lim_{\psi \to 0}\left[K^\ell(r,\psi,R)\sin\psi\right] - \tau^\ell(R,\Omega)\lim_{\psi \to 0}\left[K^\ell(r,\psi,R)\sin\psi\right] = 0 , \quad \text{(B.8.9)}$$

since we assume throughout the paper that height H of the computation point is only positive, i.e., $H \geq H_{min} > 0$, and thus $K^\ell(R + H, \psi, R) < \infty$ whenever $R > 0$. This also means that $\lim_{\psi \to 0} K^\ell(r, \psi, R) < \infty$, and thus both constituents on the right-hand side of eqn.(B.8.9) are finite.

Let us now evaluate analytically the incomplete angular integral of the spheroidal Poisson kernel occurring in eqn.(B.8.7). As $d\Omega' = \sin\psi d\psi d\alpha$, we have

$$\int_{\Omega_{\psi_0}} K^\ell(r,\psi,R)d\Omega' = \int_{\alpha=0}^{2\pi}\int_{\psi=0}^{\psi_0} K^\ell(r,\psi,R)\sin\psi d\psi d\alpha =$$

$$= 2\pi\int_{\psi=0}^{\psi_0}\left[K(r,\psi,R) - \sum_{j=0}^{\ell-1}(2j+1)\left(\frac{R}{r}\right)^{j+1}P_j(\cos\psi)\right]\sin\psi d\psi , \quad \text{(B.8.10)}$$

where we have substituted for $K^\ell(r, \psi, R)$ from eqn.(8.19). The first integral may be evaluated as follows.

$$\int_{\psi=0}^{\psi_0} K(r,\psi,R)\sin\psi d\psi = R\left(r^2 - R^2\right)\int_{\psi=0}^{\psi_0}\frac{\sin\psi d\psi}{[r^2 + R^2 - 2rR\cos\psi]^{3/2}} =$$

$$= R\left(r^2 - R^2\right)\int_{\cos\psi_0}^1\frac{dx}{[r^2 + R^2 - 2rRx]^{3/2}} = \frac{r^2 - R^2}{r}\left.\frac{1}{\sqrt{r^2 + R^2 - 2rRx}}\right|_{x=\cos\psi_0}^1 =$$

$$= \frac{r + R}{r}\left(1 - \frac{r - R}{\ell(\psi_0)}\right) , \quad \text{(B.8.11)}$$

where

$$\ell(\psi_0) := \sqrt{r^2 + R^2 - 2rR\cos\psi_0} . \quad \text{(B.8.12)}$$

In the first step, we have substituted for Poisson's kernel $K(r, \psi, R)$ from eqn.(8.20) and made the substitution $x = \cos\psi$. Then we have found the primitive function to the indefinite integral over ψ, and, finally, have substituted the lower and upper limit.

The second integral on the right-hand side of eqn.(B.8.10) reads

$$\int_{\psi=0}^{\psi_0}\sum_{j=0}^{\ell-1}(2j+1)\left(\frac{R}{r}\right)^{j+1}P_j(\cos\psi)\sin\psi d\psi =$$

$$= \frac{R}{r}\int_{\cos\psi_0}^1 dx + \sum_{j=1}^{\ell-1}(2j+1)\left(\frac{R}{r}\right)^{j+1}\int_{\cos\psi_0}^1 P_j(x)dx =$$

$$= \frac{R}{r}(1 - \cos \psi_0) - \sum_{j=1}^{\ell-1}(2j+1)\left(\frac{R}{r}\right)^{j+1} R_{j0}(\cos \psi_0) , \tag{B.8.13}$$

where $R_{jk}(x_0)$ stand for the incomplete integrals of product of two Legendre polynomials (Paul, 1973),

$$R_{jk}(x_0) := \int_{-1}^{x_0} P_j(x)P_k(x)dx . \tag{B.8.14}$$

Substituting from eqns.(B.8.10)–(B.8.13) into (B.8.7), we obtain the final form of near-zone term $\bar{\tau}^\ell_{\psi_0}(r, \Omega)$,

$$\bar{\tau}^\ell_{\psi_0}(r, \Omega) = \tau_0^\ell(r, \Omega) + \tau^\ell_{\psi_0}(r, \Omega) , \tag{B.8.15}$$

where

$$\tau_0^\ell(r, \Omega) := d^\ell(r, \psi_0, R)\tau^\ell(R, \Omega) . \tag{B.8.16}$$

The first term on the right-hand side stands for

$$d^\ell(r, \psi_0, R) := \frac{1}{2}\left[\frac{r+R}{r}\left(1 - \frac{r-R}{\ell(\psi_0)}\right) - \right.$$

$$\left. -\frac{R}{r}(1 - \cos \psi_0) + \sum_{j=1}^{\ell-1}(2j+1)\left(\frac{R}{r}\right)^{j+1} R_{j0}(\cos \psi_0)\right] , \tag{B.8.17}$$

and $\tau^\ell_{\psi_0}(r, \Omega)$ is given by eqn.(B.8.8).

B.8.2   Truncation coefficients

In analogy with Molodensky's truncation coefficients for Stokes's function (Heiskanen and Moritz, 1967, sect.7-4), we introduce the truncation coefficients for Poisson's kernel $K(r, \psi, R)$ and spheroidal Poisson's kernel $K^\ell(r, \psi, R)$.

Let us introduce an auxiliary function $K^{\psi_0}(r, \psi, R)$ as

$$K^{\psi_0}(r, \psi, R) := \begin{cases} 0 & \text{if } 0 \le \psi < \psi_0 , \\ K(r, \psi, R) & \text{if } \psi_0 \le \psi \le \pi , \end{cases} \tag{B.8.18}$$

and expand the function $K^{\psi_0}(r, \psi, R)$ into a series of Legendre polynomials,

$$K^{\psi_0}(r, \psi, R) = \sum_{j=0}^{\infty} \frac{2j+1}{2}q_j(r, \psi_0)P_j(\cos \psi) , \tag{B.8.19}$$

where $q_j(r, \psi_0)$ are expansion coefficients to be determined. Multiplying eqn.(B.8.19) by Legendre polynomial $P_k(\cos \psi)$ and integrating the result over all ψ's, we get

$$\int_{\psi=0}^{\pi} K^{\psi_0}(r, \psi, R)P_k(\cos \psi) \sin \psi d\psi =$$

$$= \sum_{j=0}^{\infty} \frac{2j+1}{2} q_j(r, \psi_0) \int_{\psi=0}^{\pi} P_j(\cos \psi) P_k(\cos \psi) \sin \psi d\psi . \tag{B.8.20}$$

Using the orthogonality property of Legendre polynomials and substituting for $K^{\psi_0}(r, \psi, R)$ from eqn.(B.8.18), the truncation coefficients $q_j(r, \psi_0)$ for Poisson's kernel $K(r, \psi, R)$ read

$$q_j(r, \psi_0) = \int_{\psi_0}^{\pi} K(r, \psi, R) P_j(\cos \psi) \sin \psi d\psi . \tag{B.8.21}$$

Provided that radius ψ_0 is chosen according to eqn.(A.8.4), the integration in eqn.(B.8.21) is taken over the far-zone domain only. In this case Poisson's kernel may be approximated by formula (A.8.5), and truncation coefficients $q_j(r, \psi_0)$ become

$$\boxed{q_j(r, \psi_0) \doteq \frac{r-R}{R} q_j(\psi_0) ,} \tag{B.8.22}$$

where

$$\boxed{q_j(\psi_0) := \int_{\psi_0}^{\pi} \frac{1}{4 \sin^3 \frac{\psi}{2}} P_j(\cos \psi) \sin \psi d\psi .} \tag{B.8.23}$$

For example,

$$q_0(\psi_0) = 2 \left(\frac{1}{\sin \frac{\psi_0}{2}} - 1 \right) . \tag{B.8.24}$$

To get a rough view into the magnitude of truncation coefficients $q_j(\psi_0)$, Figure 8.8 plots $q_j(\psi_0)$ for $\psi_0 = 1°$ and $\psi_0 = 3°$, and for $j = 0, ..., 360$.

Employing the same procedure as above, we can introduce the truncation coefficients $q_j^\ell(\psi_0)$ for the spheroidal Poisson kernel $K^\ell(r, \psi, R)$,

$$q_j^\ell(r, \psi_0) := \int_{\psi_0}^{\pi} K^\ell(r, \psi, R) P_j(\cos \psi) \sin \psi d\psi . \tag{B.8.25}$$

These coefficients can be expressed in terms of truncation coefficients $q_j(r, \psi_0)$ for Poisson's kernel $K(r, \psi, R)$ and Paul's coefficients $R_{jk}(\cos \psi_0)$ as

$$\boxed{q_j^\ell(r, \psi_0) = q_j(r, \psi_0) - \sum_{k=0}^{\ell-1} \left(\frac{R}{r} \right)^{k+1} R_{jk}(\cos \psi_0) .} \tag{B.8.26}$$

In turn, by means of the truncation coefficients $q_j^\ell(r, \psi_0)$, we can express function $K^{\ell, \psi_0}(r, \psi, R)$ which is equal to $K^\ell(r, \psi, R)$ within the interval $\psi_0 \leq \psi \leq \pi$, and vanishes elsewhere, as

$$K^{\ell, \psi_0}(r, \psi, R) = \sum_{j=0}^{\infty} \frac{2j+1}{2} q_j^\ell(r, \psi_0) P_j(\cos \psi) . \tag{B.8.27}$$

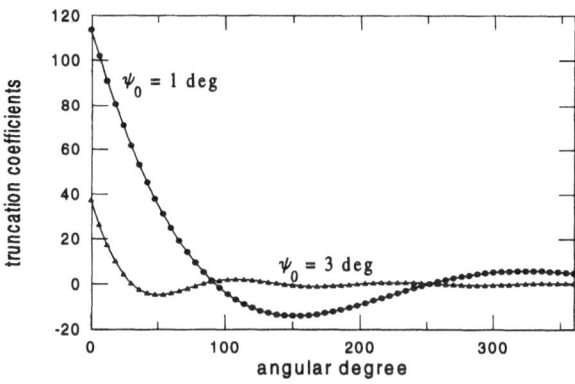

Figure 8.8: Truncation coefficients $q_j(\psi_0)$ for $\psi_0 = 1°$ and $3°$, and
$$j = 0, 1, ..., 360.$$

B.8.3 Far-zone contribution

Now, we are ready to give a spectral form of far-zone term $\tau^\ell_{\pi-\psi_0}(r, \Omega)$. Inserting
eqn.(B.8.5) and (B.8.27) into (B.8.4) and interchanging the order of integration
and summation, we obtain

$$\tau^\ell_{\pi-\psi_0}(r, \Omega) = \frac{1}{2} \sum_{j=\ell}^{j_{max}} q^\ell_j(r, \psi_0) \sum_{m=-j}^{j} \tau^\ell_{jm} Y_{jm}(\Omega) , \qquad \text{(B.8.28)}$$

where truncation coefficients $q^\ell_j(r, \psi_0)$ are given by eqn.(B.8.26).

B.8.4 Summary

Let us summarize the formulae for numerical computation of Poisson's integral
(8.17). Schematically, it may be written as a sum of three terms,

$$\boxed{\tau^\ell(r, \Omega) = \tau^\ell_0(r, \Omega) + \tau^\ell_{\psi_0}(r, \Omega) + \tau^\ell_{\pi-\psi_0}(r, \Omega) ,} \qquad \text{(B.8.29)}$$

where $\tau^\ell_0(r, \Omega)$ expresses the contribution to Poisson's integral from the integration
point being on the same geocentric radius as the computation point, eqn.(B.8.16),
$\tau^\ell_{\psi_0}(r, \Omega)$ expresses the contributions of integration points lying within the near-
zone spherical cap of radius ψ_0, eqn.(B.8.8), and $\tau^\ell_{\pi-\psi_0}(r, \Omega)$ expresses the contribution of far-zone integration points, eqn.(B.8.28).

Chapter 9

The Stokes boundary-value problem on an ellipsoid of revolution

Stokes's integral for the gravimetric determination of the geoid requires, besides other assumptions, that the gravity anomalies are referred to a sphere in geometry and gravity space. The relative error introduced by this spherical approximation is, according to Heiskanen and Moritz (1967, sect.2-14), of the order of 3×10^{-3} which causes an absolute error of 0.5 m in terms of geoidal heights. A simple error analysis reveals that this error can be even larger and that it may reach 2 m. Such an error is unacceptable at a time when the '1 cm geoid' is the target. Namely, observations of the coordinates of topographical points by means of the Global Positioning System can provide ellipsoidal heights with a relative accuracy of a few centimetres. A question arises whether the geoid can be determined with a comparable relative accuracy in order, for instance, to be able to define a system of precise orthometric heights.

To develop the theory of geoid height determination as precisely as possible is one step toward obtaining the geoid with a relative accuracy of a few centimetres. In this chapter (see also Martinec and Grafarend, 1997a), we aim to solve the Stokes boundary-value problem (**not** the Stokes two-boundary-value problem) for gravity anomalies distributed on an ellipsoid of revolution in geometry and gravity space. The ellipsoidal approximation of the geoid reflects reality much better than a spherical approximation since the actual shape of the geoid deviates from an ellipsoid of revolution by 100 m at most. Treating the geoid with respect to an ellipsoid of revolution in an appropriate normal gravity field, in particular as a boundary condition prescribed on the approximate geoid surface – the reference ellipsoid of revolution – may cause relative errors in the order of 1.5×10^{-5}. The absolute error in geoidal heights computed from the *ellipsoidal Stokes integral* does not exceed 2 mm.

9.1 Formulation of the boundary-value problem

To begin with, let us introduce ellipsoidal coordinates $\{u, \beta, \lambda\}$ through the transformation relations into the Cartesian coordinates $\{x, y, z\}$ (for details see e.g., Moon and Spencer, 1961, Table 1.07; Heiskanen and Moritz, 1967, sect.1-19; Thong and Grafarend, 1989):

$$
\begin{aligned}
x &= \sqrt{u^2 + E^2} \sin \beta \cos \lambda , \\
y &= \sqrt{u^2 + E^2} \sin \beta \sin \lambda , \\
z &= u \cos \beta ,
\end{aligned}
\tag{9.1}
$$

where $E := \sqrt{a^2 - b^2}$ (=const.) is the linear eccentricity, also called the radius of the focal circle, of the set of confocal ellipsoidal coordinate surfaces u =const. The problem that we will deal with is to determine the potential $T(u, \overline{\Omega})$, $\overline{\Omega} = (\beta, \lambda)$, on and outside the reference ellipsoid of revolution $u = b_0$ (b_0 is the minor semi-axis of the reference ellipsoid of revolution) such that

$$
\begin{aligned}
\nabla^2 T &= 0 & \text{for } u > b_0 , \tag{9.2} \\
\frac{\partial T}{\partial u} + \frac{2}{u} T &= -f & \text{for } u = b_0 , \tag{9.3} \\
T &= \frac{c}{u} + O\left(\frac{1}{u^3}\right) & \text{for } u \to \infty , \tag{9.4}
\end{aligned}
$$

where $f(\overline{\Omega})$ is assumed to be a known square-integrable function, i.e., $f(\overline{\Omega}) \in L_2(\overline{\Omega})$, and c is a constant. We will also consider a limiting case $E = 0$ when the reference ellipsoid of revolution $u = b_0$ reduces to a sphere $x^2 + y^2 + z^2 = b_0^2$, and the boundary-value problem (9.2)–(9.4) reduces to the spherical Stokes problem (Heiskanen and Moritz, 1967, sect.2-16). To guarantee the existence of a solution in this particular case, the first-degree harmonics of $f(\overline{\Omega})$ have to be removed by means of the postulate

$$
\int_{\Omega_0} f(\overline{\Omega}) Y_{1m}^*(\overline{\Omega}) d\overline{\Omega} = 0 \qquad \text{for } m = -1, 0, 1 ,
\tag{9.5}
$$

where $Y_{1m}(\overline{\Omega})$ are spherical harmonics of the first-degree and order m, the asterisk denotes a complex conjugate, Ω_0 is the full solid angle, and $d\overline{\Omega} = \sin \beta d\beta d\lambda$. Throughout this chapter we will assume that conditions (9.5) are satisfied. Moreover, to guarantee the uniqueness of the solution, the first degree harmonics have to be removed from the potential T as the asymptotic condition (9.4) indicates. The question of the uniqueness of the problem (9.2)–(9.4) for a general case $E \neq 0$ will be examined later in Section 8.5.

We will call the problem (9.2)–(9.4) the <u>ellipsoidal Stokes boundary-value problem</u> since it generalizes the traditional Stokes boundary-value problem formulated on a sphere. It approximates the boundary-value problem for geoid determination (see Chapter 1) or the Molodensky scalar boundary-value problem (Heck, 1991) by maintaining the two largest terms in boundary condition (9.3) and omitting two small ellipsoidal correction terms. (The effect of the ellipsoidal correction terms on the solution to the spherical Stokes boundary-value problem will be treated later in Chapter 11.) Alternatively, the boundary-value problem for geoid determination may be formulated in geodetic coordinates (Heck, 1991) which results in a boundary condition of a form slightly different from eqn.(9.3). We shall, however, use ellipsoidal coordinates and formulation (9.2)–(9.4) since the Laplace operator is separable in ellipsoidal coordinates (Moon and Spencer, 1961) which will substantially help us in finding a solution to the problem.

The solution to the Laplace equation (9.2) can be written in terms of ellipsoidal harmonics as follows (Moon and Spencer, 1961, p.32; Heiskanen and Moritz, 1967, sect.1-20; Thong and Grafarend, 1989):

$$T(u,\overline{\Omega}) = \sum_{j=0}^{\infty} \sum_{m=-j}^{j} T_{jm} \frac{Q_{jm}\left(i\frac{u}{E}\right)}{Q_{jm}\left(i\frac{b_0}{E}\right)} Y_{jm}(\overline{\Omega}) , \tag{9.6}$$

where $Q_{jm}\left(i\frac{u}{E}\right)$ are Legendre's function of the 2nd kind, and T_{jm} are coefficients to be determined from the boundary condition (9.3). Eqn.(A.9.11), derived in Appendix A.9, shows that

$$Q_{jm}\left(i\frac{u}{E}\right) \sim O\left(\frac{1}{u^{j+1}}\right) \qquad \text{for } u \to \infty . \tag{9.7}$$

To satisfy the asymptotic condition (9.4), the term with $j = 1$ must be excluded from the sum over j's in eqn.(9.6), i.e.,

$$T(u,\overline{\Omega}) = \sum_{\substack{j=0 \\ j \neq 1}}^{\infty} \sum_{m=-j}^{j} T_{jm} \frac{Q_{jm}\left(i\frac{u}{E}\right)}{Q_{jm}\left(i\frac{b_0}{E}\right)} Y_{jm}(\overline{\Omega}) . \tag{9.8}$$

Substituting expansion (9.8) into the boundary condition (9.3) yields

$$\sum_{\substack{j=0 \\ j \neq 1}}^{\infty} \sum_{m=-j}^{j} \frac{1}{Q_{jm}\left(i\frac{b_0}{E}\right)} \left[\left.\frac{dQ_{jm}\left(i\frac{u}{E}\right)}{du}\right|_{u=b_0} + \frac{2}{b_0} Q_{jm}\left(i\frac{b_0}{E}\right) \right] T_{jm} Y_{jm}(\overline{\Omega}) = -f(\overline{\Omega}) . \tag{9.9}$$

Moreover, expanding function $f(\overline{\Omega})$ in a series of spherical harmonics,

$$f(\overline{\Omega}) = \sum_{j=0}^{\infty} \sum_{m=-j}^{j} \int_{\Omega_0} f(\overline{\Omega}') Y_{jm}^*(\overline{\Omega}') d\overline{\Omega}' Y_{jm}(\overline{\Omega}) , \tag{9.10}$$

where the first-degree spherical harmonics of $f(\overline{\Omega})$ are equal to zero by assumption (9.5), substituting expansion (9.10) into eqn.(9.9), and comparing the coefficients at spherical harmonics $Y_{jm}(\overline{\Omega})$ in the result yields

$$T_{jm} = -\frac{\int_{\Omega_0} f(\overline{\Omega}')Y_{jm}^*(\overline{\Omega}')d\overline{\Omega}'}{\frac{1}{Q_{jm}\left(i\frac{b_0}{E}\right)}\left[\frac{dQ_{jm}\left(i\frac{u}{E}\right)}{du}\Bigg|_{u=b_0} + \frac{2}{b_0}Q_{jm}\left(i\frac{b_0}{E}\right)\right]}, \qquad (9.11)$$

for $j = 0, 2, ...,$ and $m = -j, ..., j$. Substituting coefficients T_{jm} into eqn.(9.8) and changing the order of summation over j and of integration over $\overline{\Omega}'$, due to the uniform convergence of the series expansion, the solution to the ellipsoidal Stokes boundary-value problem (9.2)–(9.4) finally reads

$$T(u,\overline{\Omega}) = \int_{\Omega_0} f(\overline{\Omega}') \sum_{\substack{j=0\\j\neq 1}}^{\infty} \sum_{m=-j}^{j} \alpha_{jm}(u)Y_{jm}^*(\overline{\Omega}')Y_{jm}(\overline{\Omega})d\overline{\Omega}', \qquad (9.12)$$

where we have introduced functions $\alpha_{jm}(u)$,

$$\alpha_{jm}(u) := -\frac{Q_{jm}\left(i\frac{u}{E}\right)}{\frac{dQ_{jm}\left(i\frac{u}{E}\right)}{du}\Bigg|_{u=b_0} + \frac{2}{b_0}Q_{jm}\left(i\frac{b_0}{E}\right)}, \qquad (9.13)$$

to abbreviate notations.

9.2 The zero-degree harmonic of T

Let us calculate the zero-degree harmonic of potential $T(u,\overline{\Omega})$. We first have:

$$\alpha_{00}(u) = -\frac{Q_{00}\left(i\frac{u}{E}\right)}{\frac{dQ_{00}\left(i\frac{u}{E}\right)}{du}\Bigg|_{u=b_0} + \frac{2}{b_0}Q_{00}\left(i\frac{b_0}{E}\right)}, \qquad (9.14)$$

where $Q_{00}\left(i\frac{u}{E}\right)$ can be expressed as (Arfken, 1968, eqn.12.222)

$$Q_{00}\left(i\frac{u}{E}\right) = -i\arctan\left(\frac{E}{u}\right) \qquad (9.15)$$

with $i = \sqrt{-1}$. Taking the derivatives of the last equation with respect to u, we get

$$\frac{dQ_{00}\left(i\dfrac{u}{E}\right)}{du} = \frac{iE}{u^2 + E^2} \ . \tag{9.16}$$

By substituting eqns.(9.15) and (9.16) into (9.14),

$$\boxed{\alpha_{00}(u) = \frac{\operatorname{arctg}\left(\dfrac{E}{u}\right)}{\dfrac{E}{b_0^2 + E^2} - \dfrac{2}{b_0}\operatorname{arctg}\left(\dfrac{E}{b_0}\right)}} \ . \tag{9.17}$$

Since $\operatorname{arctg} x \doteq x$ for $x \ll 1$, it particularly holds

$$\alpha_{00}(u) = \frac{c}{u} \qquad \text{for} \ \ u \to \infty \tag{9.18}$$

with

$$c := \frac{E}{\dfrac{E}{b_0^2 + E^2} - \dfrac{2}{b_0}\operatorname{arctg}\left(\dfrac{E}{b_0}\right)} \ . \tag{9.19}$$

9.3   Solution on the reference ellipsoid of revolution

Solution (9.12) can now be expressed as

$$T(u, \overline{\Omega}) = \frac{\alpha_{00}(u)}{4\pi} \int_{\Omega_0} f(\overline{\Omega}')d\overline{\Omega}' + \int_{\Omega_0} f(\overline{\Omega}') \sum_{j=2}^{\infty} \sum_{m=-j}^{j} \alpha_{jm}(u) Y_{jm}^{*}(\overline{\Omega}') Y_{jm}(\overline{\Omega})d\overline{\Omega}' \ . \tag{9.20}$$

In particular, we are interested in finding potential $T(u, \overline{\Omega})$ on the reference ellipsoid $u = b_0$, i.e., function $T(b_0, \overline{\Omega})$. In this case, the general formulae (9.13) and (9.20) reduce to

$$T(b_0, \overline{\Omega}) = \frac{\alpha_{00}(b_0)}{4\pi} \int_{\Omega_0} f(\overline{\Omega}')d\overline{\Omega}' + \int_{\Omega_0} f(\overline{\Omega}') \sum_{j=2}^{\infty} \sum_{m=-j}^{j} \alpha_{jm}(b_0) Y_{jm}^{*}(\overline{\Omega}') Y_{jm}(\overline{\Omega})d\overline{\Omega}' \ , \tag{9.21}$$

where $\alpha_{00}(b_0)$ are given by eqn.(9.17) for $u = b_0$ and the other $\alpha_{jm}(b_0)$ read

$$\alpha_{jm}(b_0) = -\frac{Q_{jm}\left(i\dfrac{b_0}{E}\right)}{\left.\dfrac{dQ_{jm}\left(i\dfrac{u}{E}\right)}{du}\right|_{u=b_0} + \dfrac{2}{b_0}Q_{jm}\left(i\dfrac{b_0}{E}\right)} \ . \tag{9.22}$$

From the practical point of view, the spectral form (9.21) of the solution to the Stokes boundary-value problem (9.2)–(9.4) is often inconvenient, since constructing the spectral components of $f(\overline{\Omega})$ and summing them up to high degrees and orders may become time consuming and numerically unstable. Moreover, for the case in which the reference ellipsoid of revolution $u = b_0$ deviates only slightly from a sphere, which is the case of the Earth, the solution to our problem should be close to the solution to the same problem but formulated on a sphere. We will thus attempt to rewrite $T(b_0, \overline{\Omega})$ as a sum of the well-known Stokes integral plus the corrections due to the ellipticity of the boundary. An evident advantage of such a decomposition is that existing theories as well as numerical codes for geoid height computation can simply be corrected for the ellipticity of the geoid.

To make the theory as simple as possible, but still maintain the requirements for geoidal height accuracy, throughout the following derivations, we shall retain the terms of magnitudes of the order of the first eccentricity e_0^2 of an reference ellipsoid of revolution, and neglect the term of higher-powers of e_0^2. This approximation is justifiable because the error introduced by this approximation is 1.5×10^{-5} at most which then causes an error of no more than 2 mm in the geoidal heights.

9.4 The derivative of the Legendre function of the 2nd kind

Now, let us look for the derivative of the Legendre function $Q_{jm}\left(i\dfrac{u}{E}\right)$ with respect to variable u. Equation (A.9.11), derived in Appendix A.9, shows that $Q_{jm}\left(i\dfrac{u}{E}\right)$ can be expressed as an infinite power series of the first eccentricity e,

$$e := \frac{E}{\sqrt{u^2 + E^2}} \ , \tag{9.23}$$

in the form:

$$Q_{jm}\left(i\frac{u}{E}\right) = (-1)^{m-(j+1)/2} \frac{(j+m)!}{(2j+1)!!} \, e^{j+1} \sum_{k=0}^{\infty} a_{jmk} e^{2k} \ , \tag{9.24}$$

where coefficients a_{jmk} can, for instance, be defined by recurrence relations (A.9.14). In particular, we shall need

$$a_{jm0} = 1 \ , \tag{9.25}$$

and

$$a_{jm1} = \frac{(j+1)^2 - m^2}{2(2j+3)} \ , \qquad a_{jm2} = \frac{(j+3)^2 - m^2}{4(2j+5)} \, a_{jm1} \ . \tag{9.26}$$

Throughout this monograph we assume that the eccentricity e_0 of the reference ellipsoid of revolution $u = b_0$,

$$e_0 := \frac{E}{\sqrt{b_0^2 + E^2}} \ , \tag{9.27}$$

is less than 1. For points $(u, \overline{\Omega})$ outside, or on the reference ellipsoid of revolution, i.e., when $u \geq b_0$, the series (9.24) then converges. We can take the derivative of this series with respect to u and change the order of integration and summation since the resulting series (9.28) is uniformly convergent. Consequently,

$$\frac{dQ_{jm}\left(i\frac{u}{E}\right)}{du} = (-1)^{m-(j+1)/2}\frac{(j+m)!}{(2j+1)!!}\, e^j \sum_{k=0}^{\infty}(2k+j+1)a_{jmk}e^{2k}\frac{de}{du}\,. \qquad (9.28)$$

The derivative of the eccentricity e with respect to u can easily be obtained from eqn.(9.23):

$$\frac{de}{du} = -(1-e^2)\frac{e}{u}\,. \qquad (9.29)$$

Substituting eqn.(9.29) into eqn.(9.28) yields

$$\frac{dQ_{jm}\left(i\frac{u}{E}\right)}{du} = (-1)^{m-(j+1)/2}\frac{(j+m)!}{(2j+1)!!}\,(1-e^2)\frac{e^{j+1}}{u}\sum_{k=0}^{\infty}(-2k-j-1)a_{jmk}e^{2k}\,. \qquad (9.30)$$

9.5   The uniqueness of the solution

Now, we are ready to express the sum occurring in the denominator of eqn.(9.22),

$$\left.\frac{dQ_{jm}\left(i\frac{u}{E}\right)}{du}\right|_{u=b_0} + \frac{2}{b_0}Q_{jm}\left(i\frac{b_0}{E}\right) =$$

$$= (-1)^{m-(j+1)/2}\frac{(j+m)!}{(2j+1)!!}\,\frac{e_0^{j+1}}{b_0}\sum_{k=0}^{\infty}a_{jmk}\left[(1-e_0^2)(-2k-j-1)+2\right]e_0^{2k}\,. \qquad (9.31)$$

Substituting eqns.(9.24) and (9.31) into eqn.(9.22) yields

$$\alpha_{jm}(b_0) = b_0\frac{\displaystyle\sum_{k=0}^{\infty}a_{jmk}e_0^{2k}}{\displaystyle\sum_{k=0}^{\infty}a_{jmk}\left[(1-e_0^2)(2k+j+1)-2\right]e_0^{2k}}\,. \qquad (9.32)$$

The last formula enables us to examine the uniqueness of the solution to boundary-value problem (9.2)–(9.4). Let us denote the denominator of the expression (9.32) for α_{jm} by $d_{jm}(e_0^2)$,

$$d_{jm}(e_0^2) := \sum_{k=0}^{\infty}a_{jmk}\left[(1-e_0^2)(2k+j+1)-2\right]e_0^{2k}\,, \qquad (9.33)$$

$j = 2, \cdots; \ m = 0, 1, \cdots, j$, and investigate the dependence of $d_{jm}(e_0^2)$ on e_0^2. Numerical examinations have resulted in Figure 9.1 showing the behaviour of $d_{jm}(e_0^2)$ for three values of e_0^2 as functions of angular degree j and order m, $j = 2, \cdots, 8; \ m = 0, 1, \cdots, j$; the combined index $jm := j(j+1)/2 + m + 1$. If $e_0^2 < 0.42303$, then all coefficients $d_{jm}(e_0^2)$ are positive which means that the solution to the boundary-value problem (9.2)–(9.4) is unique. Once $e_0^2 \doteq 0.42303$, $d_{22}(e_0^2) = 0$, and the solution to our problem is not unique. If the size of e_0^2 is increased further, see, for instance, the curve denoted by triangles in Figure 9.1 for $e_0^2 = 0.7$, other $d_{jm}(e_0^2)$ may then vanish and the problem (9.2)–(9.4) has a non-unique solution. We can conclude that the uniqueness of the solution can only be guaranteed if the square of the first eccentricity is less then 0.42303. Fortunately, this condition is satisfied for the Earth since $e_0^2 = 0.006\,694\,380$ (Moritz, 1980b) for the ellipsoid best-fitting the Earth's figure (the corresponding $d_{jm}(e_0^2)$ are plotted in Figure 9.1 as black dots).

9.6 The approximation up to $O(e_0^2)$

Let us arrange formula (9.32) into a form that is more suitable for highlighting Stokes's contribution. After some cumbersome but straightforward algebra, we can arrive at

$$
\alpha_{jm}(b_0) = \frac{b_0}{j-1} \left[1 - \frac{\displaystyle\sum_{k=1}^{\infty} \left[\frac{2k}{j-1} a_{jmk} - \left(1 + \frac{2k}{j-1}\right) a_{jm,k-1} \right] e_0^{2k}}{\displaystyle 1 + \sum_{k=1}^{\infty} \left(1 + \frac{2k}{j-1}\right)(a_{jmk} - a_{jm,k-1})e_0^{2k}} \right], \quad (9.34)
$$

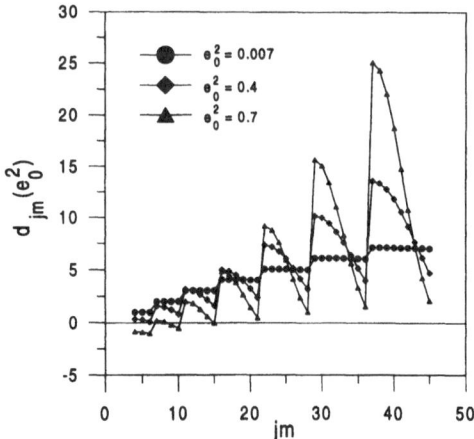

Figure 9.1: Functions $d_{jm}(e_0^2)$ for various e_0^2 as functions of degree j and order m; the combined index $jm := j(j+1)/2 + m + 1$.

where we have substituted for a_{jmo} from eqn.(9.25). The ratio of the two power series in eqn.(9.34) can further be expanded as a power series of eccentricity e_0^2. The explicit forms of the first two terms of such a series are as follows:

$$
\frac{\sum_{k=1}^{\infty} \left[\frac{2k}{j-1} a_{jmk} - \left(1 + \frac{2k}{j-1}\right) a_{jm,k-1} \right] e_0^{2k}}{1 + \sum_{k=1}^{\infty} \left(1 + \frac{2k}{j-1}\right)(a_{jmk} - a_{jm,k-1}) e_0^{2k}} = d_{jm1} e_0^2 + d_{jm2} e_0^4 + R , \qquad (9.35)
$$

where

$$
d_{jm1} := \frac{2}{j-1} a_{jm1} - \frac{j+1}{j-1} , \qquad (9.36)
$$

$$
d_{jm2} := \frac{2}{(j-1)^2} \left[2(j-1)a_{jm2} - (j+1)a_{jm1}^2 + (j+3)a_{jm1} \right] - \left(\frac{j+1}{j-1}\right)^2 , \qquad (9.37)
$$

and R is the rest of the series expansion.

We shall now attempt to estimate the sizes of particular terms on the right-hand side of eqn.(9.35). Substituting for a_{jm1} and a_{jm2} from eqn.(9.26) into eqns.(9.36) and (9.37) and after some more algebra, we get

$$
d_{jm1} = -\frac{j^2 + 3j + 2 + m^2}{(j-1)(2j+3)} , \qquad (9.38)
$$

and

$$
d_{jm2} = \frac{10j^3 + 50j^2 + 100j + 58 + 2m^2(3j+4)}{(j-1)^2(2j+3)(2j+5)} a_{jm1} - \left(\frac{j+1}{j-1}\right)^2 . \qquad (9.39)
$$

Hence

$$
|d_{jm1}| \leq \frac{2j^2 + 3j + 2}{(j-1)(2j+3)} = 1 + \frac{1}{j-1} + \frac{2}{(j-1)(2j+3)} \leq \frac{16}{7} \qquad (9.40)
$$

for $j \geq 2$ and $|m| \leq j$.

We shall continue estimating the maximum size of the first constituent creating coefficients d_{jm2}; for $j \geq 2$ and $|m| \leq j$, we have:

$$
0 < \frac{10j^3 + 50j^2 + 100j + 58 + 2m^2(3j+4)}{(j-1)^2(2j+3)(2j+5)} a_{jm1} \leq \frac{16j^3 + 58j^2 + 100j + 58}{(j-1)^2(2j+3)(2j+5)} a_{jm1}
$$

$$
\leq \frac{16j^3 + 58j^2 + 100j + 58}{(j-1)^2(2j+3)(2j+5)} a_{j01} = \frac{(j+1)^2(16j^3 + 58j^2 + 100j + 58)}{2(j-1)^2(2j+3)^2(2j+5)} <
$$

$$
< \frac{(j+1)^2(16j^3 + 58j^2 + 101j + 75)}{2(j-1)^2(2j+3)^2(2j+5)} = \frac{(j+1)^2(8j^2 + 17j + 25)}{2(j-1)^2(2j+3)(2j+5)} <
$$

$$
< \frac{(j+1)^2(8j^2 + 30j + 27)}{2(j-1)^2(2j+3)(2j+5)} = \frac{(j+1)^2(4j+9)}{2(j-1)^2(2j+5)} < \left(\frac{j+1}{j-1}\right)^2 . \qquad (9.41)
$$

Therefore, coefficients d_{jm2}, eqn.(9.39), can be estimated as

$$|d_{jm2}| < \left(\frac{j+1}{j-1}\right)^2 = \left(1 + \frac{2}{j-1}\right)^2 \leq 9 \; , \tag{9.42}$$

which is again valid for $j \geq 2$ and $|m| \leq j$.

It remains to estimate the size of the rest R of series (9.35). Since it is fairly problematic to find an analytical expression for R, we have estimated it numerically. For $e_0^2 < 1$, the power series over k in the ratio on the left-hand side of eqn.(9.35) are convergent and may be summed up. We have done it numerically for the Earth's eccentricity, $e_0^2 = 0.006\,694\,380$ (Moritz, 1980b), and for $j = 2, ..., 10^4$, and $m = 0, 1, .., j$. Note that the larger the j, the slower the decrease in the magnitude of series terms. For instance, to achieve the relative accuracy of 10^{-8} in evaluating the fraction in eqn.(9.35) for $j = 10^4$, the power series must be summed up to $k = 60$. The numerical investigations have shown that the magnitude of the rest R in eqn.(9.35) is of the order of e_0^6 at most,

$$R \sim O(e_0^6) \; . \tag{9.43}$$

Finally, retaining only the term of the order $O(e_0^2)$ and dropping out higher-order terms, eqn.(9.35) becomes

$$\frac{\sum\limits_{k=1}^{\infty} \left[\frac{2k}{j-1} a_{jmk} - \left(1 + \frac{2k}{j-1}\right) a_{jm,k-1}\right] e_0^{2k}}{1 + \sum\limits_{k=1}^{\infty} \left(1 + \frac{2k}{j-1}\right)(a_{jmk} - a_{jm,k-1})e_0^{2k}} = \left(\frac{2}{j-1} a_{jm1} - \frac{j+1}{j-1}\right) e_0^2 + O(9 \times e_0^4) \; . \tag{9.44}$$

Therefore, if we restrict ourselves only to terms of the magnitude of $O(e_0^2)$ at most, coefficients $\alpha_{jm}(b_0)$, eqn.(9.34), can then be expressed as

$$\boxed{\alpha_{jm}(b_0) \doteq \frac{b_0}{j-1}\left[1 - \left(\frac{2}{j-1} a_{jm1} - \frac{j+1}{j-1}\right) e_0^2\right] \; .} \tag{9.45}$$

Substituting eqn.(9.45) into eqn.(9.21), and bearing in mind the Laplace addition theorem for spherical harmonics (e.g., Varshalovich et al., 1989, sect.5.17.2),

$$P_j(\cos \chi) = \frac{4\pi}{2j+1} \sum_{m=-j}^{j} Y_{jm}(\overline{\Omega}) Y_{jm}^*(\overline{\Omega}') \; , \tag{9.46}$$

where $P_j(\cos \chi)$ is the Legendre polynomial of degree j, and χ is the angular distance between directions $\overline{\Omega}$ and $\overline{\Omega}'$, we get

$$\boxed{T(b_0, \overline{\Omega}) = \frac{\alpha_{00}(b_0)}{4\pi} \int_{\Omega_0} f(\overline{\Omega}') d\overline{\Omega}' + \frac{b_0}{4\pi} \int_{\Omega_0} f(\overline{\Omega}') \left[S(\chi) - e_0^2 S^{\text{ell}}(\overline{\Omega}, \overline{\Omega}')\right] d\overline{\Omega}' \; ,} \tag{9.47}$$

where $S(\chi)$ is the homogeneous spherical Stokes function (Stokes, 1849; Heiskanen and Moritz, 1967, eqn.(2-169)), and

$$S^{\text{ell}}(\overline{\Omega}, \overline{\Omega}') := 4\pi \sum_{j=2}^{\infty} \sum_{m=-j}^{j} \frac{1}{(j-1)^2} \left[\frac{(j+1)^2 - m^2}{2j+3} - (j+1) \right] Y_{jm}(\overline{\Omega}) Y_{jm}^{*}(\overline{\Omega}') .$$

(9.48)

In carrying out the last operation, we substituted for a_{jm1} from eqn.(9.26).

9.7   The ellipsoidal Stokes function

We shall call function $S^{\text{ell}}(\overline{\Omega}, \overline{\Omega}')$ the ellipsoidal Stokes function because it describes the effect of the ellipticity of boundary on the solution to Stokes's boundary-value problem. The subsequent effort will be devoted to convert the spectral form (9.48) of $S^{\text{ell}}(\overline{\Omega}, \overline{\Omega}')$ to a spatial representation.

One of the trickiest steps in such a conversion is to sum up the series

$$\sum_{m=-j}^{j} m^2 Y_{jm}(\overline{\Omega}) Y_{jm}^{*}(\overline{\Omega}') .$$

(9.49)

Thong (1993) used a simple way of finding the expression for this sum. He took the second-order derivative of the Laplace addition theorem (9.46) with respect to the longitude λ, and obtained the sum (9.49) as a linear combination of the zero-, first-, and second-order derivatives of the Legendre polynomials $P_j(\cos \chi)$. To obtain the spatial form of the Hotine function, he summed this result from $j = 2$ up to infinity. Unfortunately, this procedure leads to a strongly singular ellipsoidal Hotine function growing as $1/\chi^2$ as $\overline{\Omega}' \to \overline{\Omega}$. (Note that the original spherical Hotine function grows as $1/\chi$ as $\chi \to 0$.) Evidently, this result cannot be accepted since, in particular, as $\overline{\Omega}' \to \overline{\Omega}$, the ellipsoidal correction to the spherical Hotine problem may become larger (even for a very small eccentricity of an ellipsoid of revolution) than the contribution due to the spherical Hotine function.

In Appendix B.9, we present another way of summing up the series (9.49). Later on, we shall see that this approach ensures that the ellipsoidal Stokes function $S^{\text{ell}}(\overline{\Omega}, \overline{\Omega}')$ has the same degree of singularity at computation point $\chi = 0$ as the original spherical Stokes function $S(\chi)$. This result will be acceptable because, for an ellipsoid of revolution with a small flattening, the ellipsoidal correction to the Stokes boundary-value problem cannot become larger than the contribution due to the spherical Stokes function. Substituting eqn.(B.9.21) into eqn.(9.48), the ellipsoidal Stokes function $S^{\text{ell}}(\overline{\Omega}, \overline{\Omega}')$ can be composed from four different terms,

$$S^{\text{ell}}(\overline{\Omega}, \overline{\Omega}') =$$

$$= \sin \beta \left(\cos \beta \sin \chi \cos \chi \cos \alpha - \sin \beta \cos^2 \chi \cos^2 \alpha + \sin \beta \sin^2 \alpha \right) K_1(\cos \chi) +$$

$$+ \left(1 - \sin^2 \beta \sin^2 \alpha \right) K_2(\cos \chi) -$$

$$- \sin \beta \cos \alpha \left(\cos \beta \sin \chi - \sin \beta \cos \chi \cos \alpha \right) K_3(\cos \chi) - K_4(\cos \chi) ,$$

$$(9.50)$$

where the isotropic parts $K_i(\cos \chi)$, $i = 1, ..., 4$, of function $S^{\text{ell}}(\overline{\Omega}, \overline{\Omega}')$ are given by infinite series of the Legendre polynomials,

$$K_1(\cos \chi) := \sum_{j=3}^{\infty} \frac{2j-1}{(j-2)^2(2j+1)} \frac{dP_j(\cos \chi)}{d \cos \chi} , \tag{9.51}$$

$$K_2(\cos \chi) := \sum_{j=2}^{\infty} \frac{(j+1)^2(2j+1)}{(j-1)^2(2j+3)} P_j(\cos \chi) , \tag{9.52}$$

$$K_3(\cos \chi) := \sum_{j=3}^{\infty} \frac{j(2j-1)}{(j-2)^2(2j+1)} P_j(\cos \chi) , \tag{9.53}$$

$$K_4(\cos \chi) := \sum_{j=2}^{\infty} \frac{(j+1)(2j+1)}{(j-1)^2} P_j(\cos \chi) . \tag{9.54}$$

9.8 Spatial forms of functions $K_i(\cos \chi)$

We shall now attempt to express infinite sums for $K_i(\cos \chi)$ as finite combinations of elementary functions depending on $\cos \chi$. Following simple algebra applied to expressions (9.51)–(9.54), $K_i(\cos \chi)$ can be expressed in terms of sums

$$\sum_{j=3}^{\infty} \frac{P_j(\cos \chi)}{2j+1} \quad \text{and/or} \quad \sum_{j=3}^{\infty} \frac{P_j(\cos \chi)}{(j-1)^2} . \tag{9.55}$$

The first sum can be expressed in full elliptic integrals (Pick et al., 1973, Appendix 18) which can be calculated only approximately by a method of numerical quadrature (Press et al., 1992, sect.6.11). The second sum is equal to a definite integral, the primitive function of which cannot be expressed in closed analytical form (Pick et al., 1973, Appendix 18) but again only numerically. We can thus see that sums (9.51)–(9.54) cannot be expressed in closed analytical forms. Therefore, our method of summation will be based on the following idea. Since kernels $K_i(\cos \chi)$ are singular at point $\chi = 0$, we shall remove the contributions that are responsible for the singular behaviour at point $\chi = 0$ from sums (9.51)–(9.54). These contributions will be expressed in closed analytical forms. Having removed the singular contributions, the rest will be represented by quickly convergent infinite series, which are bounded on the whole interval $0 \leq \chi \leq \pi$. Prescribing an error of computation, they can be simply summed up numerically.

In a preparatory step, we will sum up the series of derivatives of Legendre polynomials. Making use of the recurrence relation (e.g., Arfken, 1968, eqn.(12.23))

$$\frac{dP_{j+1}(\cos\chi)}{d\cos\chi} - \frac{dP_{j-1}(\cos\chi)}{d\cos\chi} = (2j+1)P_j(\cos\chi) , \qquad (9.56)$$

we can readily derive that

$$\sum_{j=0}^{\infty}\frac{4}{(2j-1)(2j+3)}\frac{dP_j(\cos\chi)}{d\cos\chi} = \sum_{j=0}^{\infty}P_j(\cos\chi) , \qquad (9.57)$$

and

$$\sum_{j=2}^{\infty}\frac{2(4j+1)}{(j^2-1)(2j-1)(2j+3)}\frac{dP_j(\cos\chi)}{d\cos\chi} = \sum_{j=1}^{\infty}\frac{P_j(\cos\chi)}{j} + \frac{1}{10} . \qquad (9.58)$$

Furthermore, borrowing two formulae from Pick et al. (1973, eqns.(D-18;1) and (D-18;3)),

$$\sum_{j=0}^{\infty}P_j(\cos\chi) = \frac{1}{2\sin\frac{\chi}{2}} , \qquad (9.59)$$

and

$$\sum_{j=1}^{\infty}\frac{P_j(\cos\chi)}{j} = -\ln\left(\sin\frac{\chi}{2}+\sin^2\frac{\chi}{2}\right) , \qquad (9.60)$$

we get two useful formulae,

$$\sum_{j=1}^{\infty}\frac{4}{(2j-1)(2j+3)}\frac{dP_j(\cos\chi)}{d\cos\chi} = \frac{1}{2\sin\frac{\chi}{2}} , \qquad (9.61)$$

$$\sum_{j=2}^{\infty}\frac{2(4j+1)}{(j^2-1)(2j-1)(2j+3)}\frac{dP_j(\cos\chi)}{d\cos\chi} = -\ln\left(\sin\frac{\chi}{2}+\sin^2\frac{\chi}{2}\right) + \frac{1}{10} . \quad (9.62)$$

Let us start to sum up infinite series (9.51). The fraction occurring in this series can be decomposed as

$$\frac{2j-1}{(j-2)^2(2j+1)} = \frac{4}{(2j-1)(2j+3)}+$$

$$+\frac{4(4j+1)}{(j^2-1)(2j-1)(2j+3)}+\frac{3(26j^3-27j^2-18j-1)}{(j-2)^2(j^2-1)(2j-1)(2j+1)(2j+3)} \qquad (9.63)$$

In view of the last equation and eqns.(9.61) and (9.62), function $K_1(\cos\chi)$ reads

$$K_1(\cos\chi) = \frac{1}{2\sin\frac{\chi}{2}} - 2\ln\left(\sin\frac{\chi}{2}+\sin^2\frac{\chi}{2}\right) - \frac{3}{5} - \frac{16}{7}\cos\chi + R_1(\cos\chi) ,$$

$$\qquad (9.64)$$

where

$$R_1(\cos \chi) := \sum_{j=3}^{\infty} \frac{3(26j^3 - 27j^2 - 18j - 1)}{(j-2)^2(j^2-1)(2j-1)(2j+1)(2j+3)} \frac{dP_j(\cos \chi)}{d\cos \chi}. \quad (9.65)$$

Let us prove that function $R_1(\cos \chi)$ is bounded for $\chi \in \langle 0, \pi \rangle$. For these χ's, the derivatives of the Legendre polynomials can be estimated as

$$\left| \frac{dP_j(\cos \chi)}{d\cos \chi} \right| \leq j^2 , \quad (9.66)$$

which yields

$$|R_1(\cos \chi)| \leq 3 \sum_{j=3}^{\infty} \frac{j^2(26j^3 - 27j^2 - 18j - 1)}{(j-2)^2(j^2-1)(2j-1)(2j+1)(2j+3)} <$$

$$< 3 \sum_{j=3}^{\infty} \frac{j(26j^3 - 27j^2 - 18j - 1)}{(j-2)^2(j-1)(2j-1)(2j+1)(2j+3)} . \quad (9.67)$$

Using the identity

$$j(26j^3 - 27j^2 - 18j - 1) = 7j^2(2j-1)(2j+1) - 2j^4 - 27j^3 - 11j^2 - j \quad (9.68)$$

we have

$$|R_1(\cos \chi)| < 21 \sum_{j=3}^{\infty} \frac{j^2}{(j-2)^2(j-1)(2j+3)} , \quad (9.69)$$

since $2j^4 + 27j^3 + 11j^2 + j > 0$ for any $j > 0$. Finally, the sum in inequality (9.69) may be estimated as

$$\sum_{j=3}^{\infty} \frac{j^2}{(j-2)^2(j-1)(2j+3)} < \frac{1}{2} \sum_{j=1}^{\infty} \frac{j+2}{j^2(j+1)} < \frac{1}{2} \sum_{j=1}^{\infty} \left(\frac{1}{j^2} + \frac{2}{j^3} \right) = \frac{\pi^2}{12} + \zeta(3) , \quad (9.70)$$

where $\zeta(.)$ is the Riemann zeta function (e.g., Arfken, 1968, sect.5.8). In addition, we have used (Mangulis, 1965, p.50)

$$\sum_{j=1}^{\infty} \frac{1}{j^2} = \frac{\pi^2}{6} . \quad (9.71)$$

Substituting $\zeta(3) \doteq 1.20205$ (Arfken, 1968, p.359) yields

$$|R_1(\cos \chi)| < 42.8. \quad (9.72)$$

Let us point out that the last estimate is rather weak; Figure 9.3 shows that $|R_1(\cos \chi)| < 5.5$.

Nevertheless, it is important that function $R_1(\cos \chi)$ is bounded at any point in the interval $0 \leq \chi \leq \pi$, so that the singularity of function $K_1(\cos \chi)$ at point $\chi = 0$ is expressed analytically by the first two terms on the right-hand side

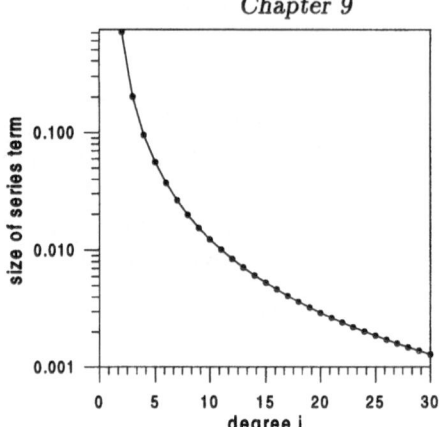

Figure 9.2: The sizes of the particular terms creating the infinite series for $R_1(1)$.

of eqn.(9.64). In addition, function $R_1(\cos \chi)$ can simply be computed numerically since it is represented by a quickly convergent series. Figure 9.2 demonstrates this fact in a transparent way; for $\chi = 0$ it plots the decay of magnitudes of series terms (9.65) with increasing degree j. Inspecting Figure 9.2 we can estimate that it is sufficient to sum up infinite series (9.65) for $R_1(\cos \chi)$ up to $j \approx 25$ in order to achieve an absolute accuracy of the order of 0.01. This accuracy is sufficient for evaluating the ellipsoidal Stokes function $S^{\mathrm{ell}}(\overline{\Omega}, \overline{\Omega}')$ in the frame of the $O(e_0^2)$-approximation.

The spatial forms of the other integral kernel $K_i(\cos \psi)$, $i = 2, 3, 4$, can be expressed in a similar fashion as the kernel $K_1(\cos \psi)$; after some cumbersome but straightforward algebra we arrive at

$$
\begin{aligned}
K_2(\cos \chi) &= \frac{1}{2 \sin \frac{\chi}{2}} - 3 \cos \chi \ln \left(\sin \frac{\chi}{2} + \sin^2 \frac{\chi}{2} \right) + 2 - 4 \cos \chi - 6 \sin \frac{\chi}{2} + \\
&\quad + R_2(\cos \chi) , \\
K_3(\cos \chi) &= -\ln \left(\sin \frac{\chi}{2} + \sin^2 \frac{\chi}{2} \right) - \cos \chi - \frac{1}{2} P_2(\cos \chi) + R_3(\cos \chi) \quad (9.73) \\
K_4(\cos \chi) &= \frac{1}{\sin \frac{\chi}{2}} - 7 \cos \chi \ln \left(\sin \frac{\chi}{2} + \sin^2 \frac{\chi}{2} \right) + 5 - 9 \cos \chi - 14 \sin \frac{\chi}{2} + \\
&\quad + R_4(\cos \chi) ,
\end{aligned}
$$

where the residuals $R_i(\cos \chi)$ are of the forms

$$
\begin{aligned}
R_2(\cos \chi) &:= \sum_{j=2}^{\infty} \frac{5j + 7}{(j-1)^2(2j+3)} P_j(\cos \chi) , \\
R_3(\cos \chi) &:= \sum_{j=3}^{\infty} \frac{6j^2 - 4j - 4}{j(j-2)^2(2j+1)} P_j(\cos \chi) , \quad\quad (9.74) \\
R_4(\cos \chi) &:= \sum_{j=2}^{\infty} \frac{6}{(j-1)^2} P_j(\cos \chi) .
\end{aligned}
$$

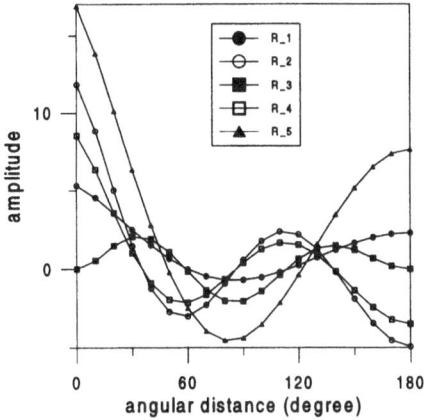

Figure 9.3: Functions $R_i(\cos\chi)$, $i = 1, ..., 4$, and $\chi \in \langle 0, \pi \rangle$.

Figure 9.3 plots the residuals $R_i(\cos\chi)$, $i = 1, ..., 4$, within the interval $0 \le \chi \le \pi$. We can see that $R_i(\cos\chi)$ are 'reasonably' smooth functions bounded for all angles χ's. This is the consequence of the fact that the magnitudes of series terms in infinite sums (9.74) decreasing quickly with increasing summation index j. In order to achieve an absolute accuracy of the order of 0.01, which is sufficient in the framework of the $O(e_0^2)$-approximation, series (9.74) may be truncated at degree of about $j \approx 25$.

Moreover, the above formulae make it possible for us to study the behaviour of functions $K_i(\cos\psi)$ in the vicinity of point $\chi = 0$. We can readily see that

$$\lim_{\chi \to 0} K_i(\cos\chi) \sim \frac{1}{\chi}, \qquad i = 1, 2$$

$$\lim_{\chi \to 0} K_3(\cos\chi) \sim -\ln\frac{\chi}{2}, \qquad (9.75)$$

$$\lim_{\chi \to 0} K_4(\cos\chi) \sim \frac{2}{\chi}.$$

Consequently, when integration point $\overline{\Omega}'$ lies in the vicinity of computation point $\overline{\Omega}$, the ellipsoidal Stokes function $S^{\mathrm{ell}}(\overline{\Omega}, \overline{\Omega}')$, see eqn.(9.50), may be approximated as

$$\boxed{S^{\mathrm{ell}}(\overline{\Omega}, \overline{\Omega}') \doteq -\frac{1}{\chi}\left(1 + \sin^2\beta\cos^2\alpha\right) \qquad \text{for } \chi \ll 1 \,.} \qquad (9.76)$$

This also means that the ellipsoidal Stokes function $S^{\mathrm{ell}}(\overline{\Omega}, \overline{\Omega}')$ has the same degree of singularity at point $\chi = 0$, namely, $1/\chi$, as the original Stokes function.

9.9 Conclusion

This chapter was motivated by the question whether the solution to the Stokes boundary-value problem with the boundary condition prescribed on an ellipsoid of revolution can be expressed in a closed spatial form, suitable for numerical computations. To answer this question, we first found the solution to the Stokes boundary-value problem in terms of ellipsoidal harmonics. The fact that this solution is represented by slowly convergent series of ellipsoidal harmonics prevents its use for regional geoid computations. That is why, in the next step, we confined ourselves to the $O(e_0^2)$-approximation, meaning that we retained terms of magnitudes of the order of e_0^2; the terms of the order of $O(e_0^4)$ and of higher powers were not considered. Nevertheless, the accuracy of the order of $O(e_0^2)$ is fairly good for today's requirement of geoid height computations. Within this accuracy, we have shown that the solution to the Stokes boundary-value problem can be expressed as an integral taken over the full solid angle, and applied to gravity anomalies multiplied by a kernel consisting of the traditional spherical Stokes function and a correction due to the elliptical shape of the boundary; we call this additional term the ellipsoidal Stokes function. We have managed to express the ellipsoidal Stokes function, originally represented in the form of an infinite sum of ellipsoidal harmonics, as a finite combination of elementary functions analytically describing the singular behaviour of the ellipsoidal Stokes function at the point $\chi = 0$. This expression is suitable for numerical solution to the Stokes boundary-value problem on an ellipsoid of revolution. The most important result is that the ellipsoidal Stokes function can be approximated by function $1/\chi$ in the vicinity of its singular point $\chi = 0$. Thus the degree of singularity of the ellipsoidal Stokes function in the vicinity of point $\chi = 0$ is the same as that of the original spherical Stokes function.

A.9   Power series expansion of the Legendre functions

For $|z| > 1$, where z is a complex number, the Legendre function of the 2nd kind, $Q_{jm}(z)$, can be defined as (Hobson, 1955, p.195, eqn.(19); Abramowitz and Stegun, 1964, eqn.(8.1.3); Gradshteyn and Ryzhik, 1980, eqn.(9.703))

$$Q_{jm}(z) = \sqrt{\pi} \frac{e^{im\pi}}{2^{j+1}} \frac{\Gamma(j+m+1)}{\Gamma\left(j+\frac{3}{2}\right)} \left(z^2 - 1\right)^{\frac{m}{2}} z^{-j-m-1} \times$$

$$\times F\left(\frac{j+m+2}{2}, \frac{j+m+1}{2}, \frac{2j+3}{2}; \frac{1}{z^2}\right) , \tag{A.9.1}$$

where $\Gamma(z)$ is the gamma function, and $F(a, b, c; z)$ is the hypergeometric function. Using the formula for the linear transformation of the hypergeometric function (Abramowitz and Stegun, 1964, eqn.(15.3.5)),

$$F(a, b, c; z) = (1 - z)^{-b} F\left(b, c - a, c; \frac{z}{z-1}\right) , \tag{A.9.2}$$

the hypergeometric function occurring in eqn.(A.9.1) becomes

$$F\left(\frac{j+m+2}{2}, \frac{j+m+1}{2}, \frac{2j+3}{2}; \frac{1}{z^2}\right) =$$

$$= \left(1 - \frac{1}{z^2}\right)^{-\frac{j+m+1}{2}} F\left(\frac{j+m+1}{2}, \frac{j-m+1}{2}, \frac{2j+3}{2}; \frac{1}{1-z^2}\right). \qquad (A.9.3)$$

Making use of the relations

$$\Gamma(j+m+1) = (j+m)!, \qquad (A.9.4)$$

$$\Gamma\left(\frac{2j+3}{2}\right) = \frac{(2j+1)!!}{2^{j+1}}\sqrt{\pi}, \qquad (A.9.5)$$

and considering eqn.(A.9.3), we have

$$Q_{jm}(z) = (-1)^m \frac{(j+m)!}{(2j+1)!!}\left(z^2 - 1\right)^{-\frac{j+1}{2}} \times$$

$$\times F\left(\frac{j+m+1}{2}, \frac{j-m+1}{2}, \frac{2j+3}{2}; \frac{1}{1-z^2}\right), \qquad (A.9.6)$$

where we have considered an integer m only, and thus $e^{im\pi} = (-1)^m$.

In particular, if

$$z := i\frac{u}{E}, \qquad (A.9.7)$$

we have

$$1 - z^2 = \frac{1}{e^2}, \qquad (A.9.8)$$

where e is the first eccentricity. Equation (A.9.6) now reads

$$Q_{jm}(i\frac{u}{E}) = (-1)^{m-(j+1)/2} \frac{(j+m)!}{(2j+1)!!} e^{j+1} F\left(\frac{j+m+1}{2}, \frac{j-m+1}{2}, \frac{2j+3}{2}; e^2\right). \qquad (A.9.9)$$

The hypergeometric function $F(a, b, c; z)$ can be expressed by the Gauss hypergeometric series (Abramowitz and Stegun, 1964, eqn.(15.1.1)):

$$F(a, b, c; z) = \frac{\Gamma(c)}{\Gamma(a)\Gamma(b)}\sum_{k=0}^{\infty} \frac{\Gamma(a+k)\Gamma(b+k)}{\Gamma(c+k)}\frac{z^k}{k!} \qquad (A.9.10)$$

for $|z| < 1$. Using the last equation, we finally arrive at

$$Q_{jm}(i\frac{u}{E}) = (-1)^{m-(j+1)/2} \frac{(j+m)!}{(2j+1)!!} e^{j+1} \sum_{k=0}^{\infty} a_{jmk} e^{2k}, \qquad (A.9.11)$$

where

$$a_{jmk} := \frac{1}{k!}\frac{\Gamma\left(\frac{j+m+1}{2}+k\right)\Gamma\left(\frac{j-m+1}{2}+k\right)\Gamma\left(\frac{2j+3}{2}\right)}{\Gamma\left(\frac{j+m+1}{2}\right)\Gamma\left(\frac{j-m+1}{2}\right)\Gamma\left(\frac{2j+3}{2}+k\right)}. \qquad (A.9.12)$$

For $k \geq 1$, coefficients a_{jmk} can be arranged in the form

$$a_{jmk} = \frac{1}{2^k \, k!} \frac{\displaystyle\prod_{\ell=0}^{k-1} \left[(j + 2\ell + 1)^2 - m^2 \right]}{\displaystyle\prod_{\ell=0}^{k-1} (2j + 2\ell + 3)} , \qquad (A.9.13)$$

which makes it possible for us to define coefficients a_{jmk} by the recurrence relation,

$$\boxed{a_{jmk} = \frac{(j + 2k - 1)^2 - m^2}{2k(2j + 2k + 1)} \, a_{jm,k-1} \qquad \text{for } k \geq 1} \qquad (A.9.14)$$

starting with $a_{jm0} = 1$.

In an analogous way, the Legendre function of the 1st kind, $P_{jm} \left(i \frac{u}{E} \right)$, can be expressed as an infinite power series of the first eccentricity e:

$$\boxed{P_{jm} \left(i \frac{u}{E} \right) = (-1)^{j/2} \frac{(2j-1)!!}{(j-m)!} \, e^{-j} \sum_{k=0}^{\infty} b_{jmk} e^{2k} } . \qquad (A.9.15)$$

The coefficients b_{jmk} are defined by the following recurrence relation

$$\boxed{b_{jmk} = \frac{(-j + 2k - 2)^2 - m^2}{2k(-2j + 2k - 1)} \, b_{jm,k-1} \qquad \text{for } k \geq 1} \qquad (A.9.16)$$

starting with $b_{jm0} = 1$.

B.9 Sum of the series (9.49)

In this section, we shall express sum (9.49) in closed form as a function of the Legendre polynomial $P_j(\cos \chi)$. The result will represent an alternative form of the addition theorem for spherical harmonics.

Let us start with the recurrence relation for spherical harmonics (Varshalovich et al., 1989, sect.5.8.2)

$$\sqrt{\frac{2j + 3}{2j + 1}} \left[(j+1)^2 - m^2 \right] Y_{jm}(\overline{\Omega}) = \sin^2 \beta \frac{\partial Y_{j+1,m}(\overline{\Omega})}{\partial \cos \beta} + (j+1) \cos \beta \, Y_{j+1,m}(\overline{\Omega}) . \tag{B.9.1}$$

Multiplying eqn.(B.9.1) by a complex conjugate recurrence relation to (B.9.1) taken at point $\overline{\Omega}'$, we get

$$\frac{2j + 3}{2j + 1} \left[(j+1)^2 - m^2 \right] Y_{jm}(\overline{\Omega}) Y_{jm}^*(\overline{\Omega}') =$$

$$= \sin^2 \beta \sin^2 \beta' \frac{\partial Y_{j+1,m}(\overline{\Omega})}{\partial \cos \beta} \frac{\partial Y^*_{j+1,m}(\overline{\Omega}')}{\partial \cos \beta'} +$$

$$+ (j+1) \cos \beta \sin^2 \beta' Y_{j+1,m}(\overline{\Omega}) \frac{\partial Y^*_{j+1,m}(\overline{\Omega}')}{\partial \cos \beta'} + \qquad (B.9.2)$$

$$+ (j+1) \cos \beta' \sin^2 \beta \, Y^*_{j+1,m}(\overline{\Omega}') \frac{\partial Y_{j+1,m}(\overline{\Omega})}{\partial \cos \beta} +$$

$$+ (j+1)^2 \cos \beta \cos \beta' Y_{j+1,m}(\overline{\Omega}) \, Y^*_{j+1,m}(\overline{\Omega}') \, .$$

Summing eqn.(B.9.2) from $m = -j - 1$ up to $m = j + 1$, and realizing that the factor $(j+1)^2 - m^2$ is equal to zero for $m = \pm(j+1)$, we get

$$\frac{2j+3}{2j+1} \sum_{m=-j}^{j} \left[(j+1)^2 - m^2 \right] Y_{jm}(\overline{\Omega}) Y^*_{jm}(\overline{\Omega}') =$$

$$= \sin^2 \beta \sin^2 \beta' \sum_{m=-(j+1)}^{j+1} \frac{\partial Y_{j+1,m}(\overline{\Omega})}{\partial \cos \beta} \frac{\partial Y^*_{j+1,m}(\overline{\Omega}')}{\partial \cos \beta'} +$$

$$+ (j+1) \cos \beta \sin^2 \beta' \sum_{m=-(j+1)}^{j+1} Y_{j+1,m}(\overline{\Omega}) \frac{\partial Y^*_{j+1,m}(\overline{\Omega}')}{\partial \cos \beta'} + \qquad (B.9.3)$$

$$+ (j+1) \cos \beta' \sin^2 \beta \sum_{m=-(j+1)}^{j+1} Y^*_{j+1,m}(\overline{\Omega}') \frac{\partial Y_{j+1,m}(\overline{\Omega})}{\partial \cos \beta} +$$

$$+ (j+1)^2 \cos \beta \cos \beta' \sum_{m=-(j+1)}^{j+1} Y_{j+1,m}(\overline{\Omega}) \, Y^*_{j+1,m}(\overline{\Omega}') \, .$$

The sums of the products of spherical harmonics and their derivatives will be simplified by means of the Laplace addition theorem for spherical harmonics. Taking the addition theorem (9.46) for index $j + 1$, we have

$$\frac{2j+3}{4\pi} P_{j+1}(\cos \chi) = \sum_{m=-(j+1)}^{j+1} Y_{j+1,m}(\overline{\Omega}) \, Y^*_{j+1,m}(\overline{\Omega}') \, , \qquad (B.9.4)$$

where χ is the angular distance between points $\overline{\Omega}$ and $\overline{\Omega}'$. Differentiating addition theorem (B.9.4) with respect to $\cos \beta$ and $\cos \beta'$, and using the results in eqn.(B.9.3) yields

$$\frac{4\pi}{2j+1} \sum_{m=-j}^{j} \left[(j+1)^2 - m^2 \right] Y_{jm}(\overline{\Omega}) Y^*_{jm}(\overline{\Omega}') =$$

$$= \sin^2 \beta \sin^2 \beta' \frac{\partial^2 P_{j+1}(\cos \chi)}{\partial \cos \beta \, \partial \cos \beta'} + (j+1) \cos \beta \sin^2 \beta' \frac{\partial P_{j+1}(\cos \chi)}{\partial \cos \beta'} + \qquad (B.9.5)$$

$$+ (j+1) \cos \beta' \sin^2 \beta \frac{\partial P_{j+1}(\cos \chi)}{\partial \cos \beta} + (j+1)^2 \cos \beta \cos \beta' P_{j+1}(\cos \chi) \, .$$

The partial derivatives of Legendre polynomials $P_{j+1}(\cos \chi)$ with respect to $\cos \beta$ and $\cos \beta'$ will now be expressed in terms of the ordinary derivatives of the Legendre polynomials with respect to $\cos \chi$ by making use of the well-known formulae of spherical trigonometry (e.g., Heiskanen and Moritz, 1967, eqn.(2-208)),

$$\cos \chi = \cos \beta \cos \beta' + \sin \beta \sin \beta' \cos(\lambda - \lambda') , \qquad (B.9.6)$$
$$\sin \chi \cos \alpha = \sin \beta \cos \beta' - \cos \beta \sin \beta' \cos(\lambda - \lambda') , \qquad (B.9.7)$$
$$\sin \chi \sin \alpha = -\sin \beta' \sin(\lambda - \lambda') , \qquad (B.9.8)$$

where α is the azimuth between directions $\overline{\Omega}$ and $\overline{\Omega}'$. Realizing that

$$\frac{\partial P_{j+1}(\cos \chi)}{\partial \cos \beta} = \frac{dP_{j+1}(\cos \chi)}{d \cos \chi} \frac{\partial \cos \chi}{\partial \cos \beta} , \qquad (B.9.9)$$

and taking the derivative of eqn.(B.9.6) with respect to $\cos \beta$, we have

$$\frac{\partial P_{j+1}(\cos \chi)}{\partial \cos \beta} = [\cos \beta' - \cot \beta \sin \beta' \cos(\lambda - \lambda')] \frac{dP_{j+1}(\cos \chi)}{d \cos \chi} . \qquad (B.9.10)$$

Similarly,

$$\frac{\partial P_{j+1}(\cos \chi)}{\partial \cos \beta'} = [\cos \beta - \cot \beta' \sin \beta \cos(\lambda - \lambda')] \frac{dP_{j+1}(\cos \chi)}{d \cos \chi} . \qquad (B.9.11)$$

Furthermore, differentiating eqn.(B.9.10) with respect to $\cos \beta'$, we obtain the 2nd-order derivative occurring in eqn.(B.9.5),

$$\frac{\partial^2 P_{j+1}(\cos \chi)}{\partial \cos \beta \, \partial \cos \beta'} = [1 + \cot \beta \cot \beta' \cos(\lambda - \lambda')] \frac{dP_{j+1}(\cos \chi)}{d \cos \chi} +$$

$$+ [\cos \beta' - \cot \beta \sin \beta' \cos(\lambda - \lambda')] [\cos \beta - \cot \beta' \sin \beta \cos(\lambda - \lambda')] \frac{d^2 P_{j+1}(\cos \chi)}{d(\cos \chi)^2} . \qquad (B.9.12)$$

Substituting eqns.(B.9.10)–(B.9.12) into eqn.(B.9.5) yields

$$\frac{4\pi}{2j+1} \sum_{m=-j}^{j} \left[(j+1)^2 - m^2 \right] Y_{jm}(\overline{\Omega}) Y_{jm}^*(\overline{\Omega}') =$$

$$= \sin \beta \sin \beta' [\sin \beta \cos \beta' - \cos \beta \sin \beta' \cos(\lambda - \lambda')] \times$$

$$\times [\sin \beta' \cos \beta - \cos \beta' \sin \beta \cos(\lambda - \lambda')] \frac{d^2 P_{j+1}(\cos \chi)}{d(\cos \chi)^2} + \qquad (B.9.13)$$

$$+ \sin \beta \sin \beta' [\sin \beta \sin \beta' + \cos \beta \cos \beta' \cos(\lambda - \lambda')] \frac{dP_{j+1}(\cos \chi)}{d \cos \chi} +$$

$$+ (j+1) \left[\cos^2 \beta \sin^2 \beta' - 2 \sin \beta \cos \beta \sin \beta' \cos \beta' \cos(\lambda - \lambda') + \right.$$

$$+ \left. \sin^2 \beta \cos^2 \beta' \right] \frac{dP_{j+1}(\cos \chi)}{d \cos \chi} + (j+1)^2 \cos \beta \cos \beta' P_{j+1}(\cos \chi) .$$

The functions in square brackets standing in front of the derivatives of the Legendre polynomials will be expressed in terms of angular distance χ and the azimuth α between directions $\overline{\Omega}$ and $\overline{\Omega}'$. Using eqns.(B.9.6)–(B.9.8), after some we obtain:

$$\sin \beta' \cos \beta - \cos \beta' \sin \beta \cos(\lambda - \lambda') = \frac{1}{\sin \beta'} (\cos \beta - \cos \beta' \cos \chi) , \quad \text{(B.9.14)}$$

$$\sin \beta \sin \beta' + \cos \beta \cos \beta' \cos(\lambda - \lambda') = \frac{1}{\sin \beta'} (\sin \beta - \cos \beta' \sin \chi \cos \alpha) , \quad \text{(B.9.15)}$$

$$\cos^2 \beta \sin^2 \beta' - 2 \sin \beta \cos \beta \sin \beta' \cos \beta' \cos(\lambda - \lambda') + \sin^2 \beta \cos^2 \beta' =$$
$$= \sin^2 \chi \left(1 - \sin^2 \beta \sin^2 \alpha\right) . \quad \text{(B.9.16)}$$

Substituting eqns.(B.9.14)–(B.9.16) into eqn.(B.9.13), and employing Legendre's differential equation,

$$\sin^2 \chi \frac{d^2 P_{j+1}(\cos \chi)}{d(\cos \chi)^2} - 2 \cos \chi \frac{dP_{j+1}(\cos \chi)}{d \cos \chi} + (j+1)(j+2)P_{j+1}(\cos \chi) = 0 , \quad \text{(B.9.17)}$$

we obtain

$$\frac{4\pi}{2j+1} \sum_{m=-j}^{j} \left[(j+1)^2 - m^2\right] Y_{jm}(\overline{\Omega})Y_{jm}^*(\overline{\Omega}') =$$

$$= \frac{\sin \beta \cos \alpha}{\sin \chi} (\cos \beta - \cos \beta' \cos \chi) \left[2 \cos \chi \frac{dP_{j+1}(\cos \chi)}{d \cos \chi} - \quad \text{(B.9.18)}\right.$$

$$-(j+1)(j+2)P_{j+1}(\cos \chi)] + \sin \beta (\sin \beta - \cos \beta' \sin \chi \cos \alpha) \frac{dP_{j+1}(\cos \chi)}{d \cos \chi} +$$

$$+(j+1)\sin^2 \chi \left(1 - \sin^2 \beta \sin^2 \alpha\right) \frac{dP_{j+1}(\cos \chi)}{d \cos \chi} + (j+1)^2 \cos \beta \cos \beta' P_{j+1}(\cos \chi) .$$

Multiplying eqn.(B.9.6) by $\cos \beta / \sin \beta$ and adding the result to eqn.(B.9.7) yields

$$\cos \beta' = \cos \beta \cos \chi + \sin \beta \sin \chi \cos \alpha . \quad \text{(B.9.19)}$$

By employing the last equation and the recurrence formula for the derivative of Legendre polynomials,

$$\sin^2 \chi \frac{dP_{j+1}(\cos \chi)}{d \cos \chi} + (j+1) \cos \chi P_{j+1}(\cos \chi) = (j+1)P_j(\cos \chi) , \quad \text{(B.9.20)}$$

in eqn.(B.9.18), we finally have

$$\frac{4\pi}{2j+1} \sum_{m=-j}^{j} \left[(j+1)^2 - m^2\right] Y_{jm}(\overline{\Omega})Y_{jm}^*(\overline{\Omega}') =$$

$$= \sin \beta \left(\cos \beta \sin \chi \cos \chi \cos \alpha - \sin \beta \cos^2 \chi \cos^2 \alpha + \sin \beta \sin^2 \alpha\right) \frac{dP_{j+1}(\cos \chi)}{d \cos \chi} -$$

$$-(j+1)\sin \beta \cos \alpha (\cos \beta \sin \chi - \sin \beta \cos \chi \cos \alpha) P_{j+1}(\cos \chi) + \quad \text{(B.9.21)}$$

$$+(j+1)^2 \left(1 - \sin^2 \beta \sin^2 \alpha\right) P_j(\cos \chi) .$$

In particular, when $\overline{\Omega} = \overline{\Omega}'$, we get

$$\frac{4\pi}{2j+1} \sum_{m=-j}^{j} \left[(j+1)^2 - m^2\right] \left|Y_{jm}(\overline{\Omega})\right|^2 = \frac{1}{2}(j+1)(j+2)\sin^2\beta + (j+1)^2\cos^2\beta \ .$$

$$(\text{B.9.22})$$

Chapter 10

The external Dirichlet boundary-value problem for the Laplace equation on an ellipsoid of revolution

To be able to work with a harmonic gravitational potential outside the geoid, it is necessary to replace the gravitational effect of the topographical masses by the gravitational effect of an auxiliary body situated below or on the geoid. To carry out this 'mass-redistribution' mathematically, a number more or less idealized so-called compensation models have been proposed. Although the way of compensation might originally be motivated by geophysical ideas, for the purpose of geoid computation, we may, in principle, employ any compensation model generating a harmonic gravitational field outside the geoid. For instance, the topographic-isostatic compensation models (Section 3.3) are based on the anomalies of density distribution in a layer between the geoid and the compensation level. In the limiting case, the topographical masses may be compensated by a mass surface located on the geoid, i.e., by a layer whose thickness is infinitely small. This kind of compensation is called Helmert's 2nd condensation (Section 3.4).

Redistributing the topographical masses and subtracting the centrifugal potential from the 'observed' surface geopotential, we get the gravitational potential induced by the masses below the geoid, which is harmonic in the space above the geoid. To find it on the geoid, a fixed boundary-value problem governed by the Laplace equation with the Dirichlet boundary condition on the Earth's surface (determined by GPS observations) can be formulated.

The problem can be solved by employing Poisson's integral provided that the geoid in the boundary condition for a potential is approximated by a mean sphere (Section 8.2). Such an approximation may produce significant errors in geoidal heights. The only possibility to reduce them is to approximate the geoid by a surface fitting it in a closer way. An ellipsoid of revolution, best-fitting the geoid, represents an alternative to approximate the geoid closer to the reality. This motivates the question: Can Green's function to the external Dirichlet boundary-value problem for the Laplace equation with data distributed on an ellipsoid of revolution be constructed? We have already constructed Green's function to the Stokes problem for gravity data distributed on an ellipsoid of revolution (Chapter 9). In this chapter, we intend to apply the same analytical approach to solve the raised question (see also Martinec and Grafarend, 1997b).

The construction of this Green's function (later on, we called it the ellipsoidal Poisson kernel) enables us to avoid making the assumption on a spherical approximation of the geoid in the problem of the upward and downward continuation of gravitation. Even pessimistic estimates show that ellipsoidal approximation of the geoid in this type of the boundary-value problem is sufficient to determine a very accurate geoid with an error not exceeding 1 cm. Such an accuracy is the ultimate goal of scientists dealing with the determination of a very precise geoid.

10.1   Formulation of the boundary-value problem

The problem we will deal with is to determine a function $T(u, \overline{\Omega})$, $\overline{\Omega} := (\beta, \lambda)$, *outside* the reference ellipsoid of revolution $u = b_0$ such that

$$
\begin{aligned}
\nabla^2 T &= 0 & &\text{for } u > b_0 \ , & (10.1)\\
T &= f & &\text{for } u = b_0 \ , & (10.2)\\
T &\sim O\left(\frac{1}{u}\right) & &\text{for } u \to \infty \ , & (10.3)
\end{aligned}
$$

where (u, β, λ) are ellipsoidal coordinates defined by eqn.(9.1), $f(\overline{\Omega})$ is assumed to be a known square-integrable function, $f(\overline{\Omega}) \in L_2(\overline{\Omega})$. The asymptotic condition (10.3) means that the harmonic function T approaches zero at infinity.

The solution to the Laplace equation (10.1) can be written in terms of ellipsoidal harmonics as shown by eqn.(9.6):

$$
T(u, \overline{\Omega}) = \sum_{j=0}^{\infty} \sum_{m=-j}^{j} T_{jm} \frac{Q_{jm}\left(i\dfrac{u}{E}\right)}{Q_{jm}\left(i\dfrac{b_0}{E}\right)} Y_{jm}(\overline{\Omega}) \ , \tag{10.4}
$$

where $Q_{jm}\left(i\dfrac{u}{E}\right)$ are Legendre's function of the 2nd kind, $Y_{jm}(\overline{\Omega})$ are complex spherical harmonics of degree j and order m, and T_{jm} are coefficients to be determined from the boundary condition (10.2).

Substituting eqn.(10.4) into eqn.(10.2), expanding the function $f(\overline{\Omega})$ in a series of spherical harmonics,

$$
f(\overline{\Omega}) = \sum_{j=0}^{\infty} \sum_{m=-j}^{j} \int_{\Omega_0} f(\overline{\Omega}') Y_{jm}^*(\overline{\Omega}') d\overline{\Omega}' Y_{jm}(\overline{\Omega}) \ , \tag{10.5}
$$

where Ω_0 is the full solid angle and $d\overline{\Omega} = \sin\beta d\beta d\lambda$, and comparing the coefficients at spherical harmonics $Y_{jm}(\overline{\Omega})$ in the result, we get

$$T_{jm} = \int_{\Omega_0} f(\overline{\Omega}')Y_{jm}^*(\overline{\Omega}')d\overline{\Omega}' \tag{10.6}$$

for $j = 0, 1, ...,$ and $m = -j, ..., j$. Furthermore, substituting coefficients T_{jm} into
eqn.(10.4), interchanging the order of summation over j and m and integration
over $\overline{\Omega}'$ due to the uniform convergence of the series expansion (10.4), the solution
to the ellipsoidal Dirichlet boundary-value problem (10.1)–(10.3) reads

$$T(u,\overline{\Omega}) = \int_{\Omega_0} f(\overline{\Omega}')\sum_{j=0}^{\infty}\sum_{m=-j}^{j} \frac{Q_{jm}\left(i\dfrac{u}{E}\right)}{Q_{jm}\left(i\dfrac{b_0}{E}\right)} Y_{jm}^*(\overline{\Omega}')Y_{jm}(\overline{\Omega})d\overline{\Omega}' \ . \tag{10.7}$$

From the practical point of view, the spectral form (10.7) of the solution to
the Dirichlet problem (10.1)–(10.3) may often become inconvenient, since the
computation of $Q_{jm}(z)$ functions and their summation up to high degrees and
orders $(j \sim 10^4 - 10^5)$ is time consuming and numerically unstable (Sona, 1995).
Moreover, in the case that the reference ellipsoid of revolution $u = b_0$ deviates
from a sphere by only a tiny amount which is the case for the Earth, the solution
to our problem should be close to the solution to the same problem but formulated
on a sphere. We will thus attempt to rewrite $T(u,\overline{\Omega})$ as a sum of the well-known
Poisson integral (Section 7.2) that solves the Dirichlet problem on a sphere, plus
the corrections due to the ellipticity of the boundary. An evident advantage
of such a decomposition is that existing theories as well as numerical codes for
solving the Dirichlet problem on a sphere can simply be corrected for the ellipticity
of the boundary.

10.2 Power series representation of the integral kernel

By eqn.(A.9.11), the ratio of the Legendre functions of the second kind occurring
in eqn.(10.7) reads

$$\frac{Q_{jm}\left(i\dfrac{u}{E}\right)}{Q_{jm}\left(i\dfrac{b_0}{E}\right)} = \left(\frac{e}{e_0}\right)^{j+1} \frac{\displaystyle\sum_{k=0}^{\infty} a_{jmk}e^{2k}}{\displaystyle\sum_{k=0}^{\infty} a_{jmk}e_0^{2k}} \ , \tag{10.8}$$

where a_{jmk} are given by eqn.(A.9.14). Dividing the polynomials in eqn.(10.8) term by term, we can write

$$\frac{\sum\limits_{k=0}^{\infty} a_{jmk} e^{2k}}{\sum\limits_{k=0}^{\infty} a_{jmk} e_0^{2k}} = 1 + \sum\limits_{k=1}^{\infty} b_{jmk} , \tag{10.9}$$

where the explicit forms of the first few constituents read

$$b_{jm1} = a_{jm1}\left(e^2 - e_0^2\right) , \tag{10.10}$$

$$b_{jm2} = a_{jm2}\left(e^4 - e_0^4\right) - a_{jm1}^2 e_0^2\left(e^2 - e_0^2\right) , \tag{10.11}$$

$$b_{jm3} = a_{jm3}\left(e^6 - e_0^6\right) - a_{jm2}a_{jm1}e_0^2\left(e^4 + e^2 e_0^2 - 2e_0^4\right) + $$
$$+ a_{jm1}^3 e_0^4\left(e^2 - e_0^2\right) . \tag{10.12}$$

Generally,

$$b_{jmk} \sim O\left(e^{2r} e_0^{2s}\right) , \qquad r + s = k . \tag{10.13}$$

To get an analytical expression for the k-th term of series (10.9) means the performance of some cumbersome algebraic manipulations. To avoid them, we will confine ourselves to the case where the computation point ranges in a limited layer above the reference ellipsoid of revolution (e.g., topographical layer), namely, $b_0 < u < b_0 + 9000$ m, which includes all the actual topographical masses of the Earth. For this restricted case, which is, however, often considered when geodetic boundary-value problems are solved, let us express the first eccentricity e of the computation point by means of the first eccentricity e_0 of the reference ellipsoid of revolution and a quantity ε, $\varepsilon > 0$, as

$$e = e_0(1 - \varepsilon) . \tag{10.14}$$

Assuming $b_0 < u < b_0 + 9000$ m means that $\varepsilon < 1.4 \times 10^{-3}$, and we can put approximately

$$e^{2k} \doteq e_0^{2k}\left(1 - 2k\varepsilon\right) , \tag{10.15}$$

$$b_{jm1} \doteq -2\varepsilon e_0^2 a_{jm1} . \tag{10.16}$$

This allows us to write

$$\frac{\sum\limits_{k=0}^{\infty} a_{jmk} e^{2k}}{\sum\limits_{k=0}^{\infty} a_{jmk} e_0^{2k}} = 1 - 2\varepsilon \frac{\sum\limits_{k=1}^{\infty} k\, a_{jmk} e_0^{2k}}{\sum\limits_{k=0}^{\infty} a_{jmk} e_0^{2k}} . \tag{10.17}$$

Let us expand the fraction on the right-hand side of the last equation (divided by a_{jm1}) into an infinite power series of e_0^2,

$$\frac{1}{a_{jm1}} \frac{\sum\limits_{k=1}^{\infty} k \, a_{jmk} e_0^{2k}}{\sum\limits_{k=0}^{\infty} a_{jmk} e_0^{2k}} = \sum_{k=1}^{\infty} \beta_{jmk} e_0^{2k} \ . \tag{10.18}$$

To find the coefficients β_{jmk}, let us rewrite the last equation in the form

$$\frac{1}{a_{jm1}} \sum_{k=1}^{\infty} k \, a_{jmk} e_0^{2k} = \sum_{k=0}^{\infty} a_{jmk} e_0^{2k} \sum_{\ell=1}^{\infty} \beta_{jm\ell} e_0^{2\ell} \ . \tag{10.19}$$

Since both the series on the right-hand side are absolute convergent, their product may be rearranged as

$$\sum_{k=0}^{\infty} a_{jmk} e_0^{2k} \sum_{\ell=1}^{\infty} \beta_{jm\ell} e_0^{2\ell} = \sum_{k=1}^{\infty} \sum_{\ell=1}^{k} \beta_{jm\ell} a_{jm,k-\ell} e_0^{2k} \ . \tag{10.20}$$

Substituting eqn.(10.20) into eqn.(10.19) and equating coefficients at e_0^{2k} on both sides of the result, we obtain

$$k \frac{a_{jmk}}{a_{jm1}} = \sum_{\ell=1}^{k} \beta_{jm\ell} a_{jm,k-\ell} \qquad k = 1, 2, \dots , \tag{10.21}$$

which yields the recurrence relation for β_{jmk}:

$$\boxed{\beta_{jmk} = k \frac{a_{jmk}}{a_{jm1}} - \sum_{\ell=1}^{k-1} \beta_{jm\ell} a_{jm,k} \, \ell \qquad k = 2, 3, \dots} \tag{10.22}$$

with the starting value

$$\beta_{jm1} = 1 \ . \tag{10.23}$$

With the recurrence relation (10.22) we may easily construct the higher coefficients β_{jmk}:

$$\beta_{jm2} = 2 \frac{a_{jm2}}{a_{jm1}} - a_{jm1} \ , \tag{10.24}$$

$$\beta_{jm3} = 3 \frac{a_{jm3}}{a_{jm1}} - 3 a_{jm2} + a_{jm1}^2 \ . \tag{10.25}$$

This process may be continued indefinitely.

Important properties of coefficients β_{jmk} are:

$$1 = \beta_{jm1} > \beta_{jm2} > \beta_{jm3} > \dots \geq 0 \ , \tag{10.26}$$

$$\beta_{jjk} > \beta_{j,j-1,k} > \dots > \beta_{j0k} \ . \tag{10.27}$$

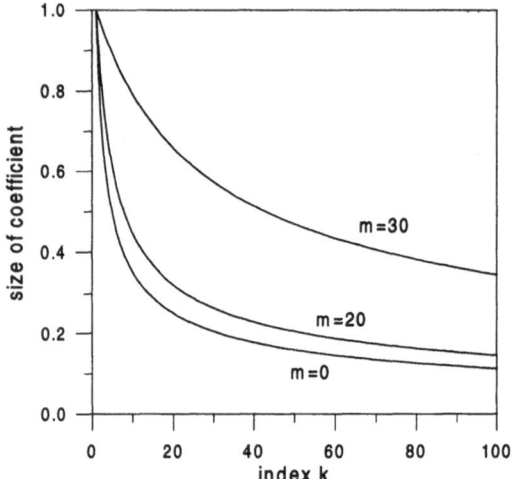

Figure 10.1: Coefficients β_{jmk} for $j = 30$, $m = 0, 20, 30$, and $k = 1, ..., 100$.

Figure 10.1 demonstrates these properties for $j = 30$.

Finally, substituting series (10.18) into eqn.(10.17) and using eqns.(10.16) and (10.23) for b_{jm1} and β_{jm1}, respectively, yields

$$\frac{\sum\limits_{k=0}^{\infty} a_{jmk} e^{2k}}{\sum\limits_{k=0}^{\infty} a_{jmk} e_0^{2k}} = 1 + b_{jm1} \left(1 + \sum\limits_{k=2}^{\infty} \beta_{jmk} e_0^{2k-2} \right) . \tag{10.28}$$

10.3   The approximation up to $O(e_0^2)$

The harmonic upward or downward continuation of the potential or the gravitation between the geoid to the Earth's surface is an example of practical application of the boundary-value problem (10.1)–(10.3). To make the theory as simple as possible but still matching the requirements on geoid height accuracy, we will keep throughout following derivations the terms of magnitudes of the order of the first eccentricity e_0^2 of the Earth's reference ellipsoid of revolution, and neglect the term of higher-powers of e_0^2. This approximation is justifiable because the error introduced by this approximation is 1.5×10^{-5} at most, which then causes an error of at most 2 mm in the geoidal heights.

Keeping in mind the inequalities (10.26), and assuming that $e_0^2 \ll 1$, the magnitude of the last term in eqn.(10.28) is of the order of e_0^2,

$$\sum\limits_{k=2}^{\infty} \beta_{jmk} e_0^{2k-2} \sim O\left(e_0^2\right) . \tag{10.29}$$

Hence, eqn.(10.8) becomes

$$\frac{Q_{jm}\left(i\frac{u}{E}\right)}{Q_{jm}\left(i\frac{b_0}{E}\right)} = \left(\frac{e}{e_0}\right)^{j+1}\left(1 + a_{jm1}\left(e^2 - e_0^2\right)\left[1 + O\left(e_0^2\right)\right]\right). \tag{10.30}$$

Substituting eqn.(10.30) into eqn.(10.7), evaluating a_{jm1} according to eqn.(A.9.14) for $k = 1$, bearing in mind the Laplace addition theorem (9.46) for spherical harmonics, we obtain

$$T(u,\overline{\Omega}) = \frac{1}{4\pi}\int_{\Omega_0} f(\overline{\Omega}')\left[K(t,\cos\chi) - e_0^2 k^{\text{ell}}(t,\overline{\Omega},\overline{\Omega}')\left(1 + O\left(e_0^2\right)\right)\right]d\overline{\Omega}', \tag{10.31}$$

where χ is the angular distance between directions $\overline{\Omega}$ and $\overline{\Omega}'$,

$$t := \frac{e}{e_0}, \tag{10.32}$$

$K(t,\cos\chi)$ is the spherical Poisson kernel (Kellogg, 1929, sect.9.4),

$$K(t,\cos\chi) = \sum_{j=0}^{\infty}(2j+1)t^{j+1}P_j(\cos\chi) = \frac{t(1-t^2)}{g^3}, \tag{10.33}$$

with

$$g \equiv g(t,\cos\chi) := \sqrt{1 + t^2 - 2t\cos\chi}, \tag{10.34}$$

and $k^{\text{ell}}(t,\overline{\Omega},\overline{\Omega}')$ stands for

$$\boxed{k^{\text{ell}}(t,\overline{\Omega},\overline{\Omega}') := 4\pi\left(1 - t^2\right)\sum_{j=0}^{\infty}\sum_{m=-j}^{j}t^{j+1}\frac{(j+1)^2 - m^2}{2(2j+3)}Y_{jm}(\overline{\Omega})Y_{jm}^*(\overline{\Omega}').} \tag{10.35}$$

Moreover, it holds ($t < 1$):

$$\left|k^{\text{ell}}(t,\overline{\Omega},\overline{\Omega}')\right| \le k^{\text{ell}}(t,\overline{\Omega},\overline{\Omega}) \le k^{\text{ell}}(t,\overline{\Omega},\overline{\Omega})\Big|_{\vartheta=0} = \frac{1-t^2}{2}\sum_{j=0}^{\infty}t^{j+1}(j+1)^2\frac{2j+1}{2j+3} <$$

$$< \frac{1-t^2}{2}\sum_{j=0}^{\infty}t^{j+1}(j+1)^2 = \frac{t(1+t)(1-t^2)}{2(1-t)^3} < \frac{t(1-t)^2}{(1-t)^3} = K(t,1), \tag{10.36}$$

where we have used the relations:

$$Y_{jm}(0,\lambda) = \delta_{m0}\sqrt{\frac{2j+1}{4\pi}}, \tag{10.37}$$

and

$$\sum_{j=0}^{\infty}t^j(j+1)^2 = \frac{1+t}{(1-t)^3}, \tag{10.38}$$

borrowed from Varshalovich et al. (1989, sect.5.14) and Mangulis (1965, p.72), respectively.

The estimate (10.36) helps us to restrict ourselves to the terms of the magnitude of $O(e_0^2)$ at most. Within this accuracy, eqn.(10.31) reads

$$T(u, \overline{\Omega}) = \frac{1}{4\pi} \int_{\Omega_0} f(\overline{\Omega}') K^{\text{ell}}(t, \overline{\Omega}, \overline{\Omega}') d\overline{\Omega}' , \qquad (10.39)$$

where

$$K^{\text{ell}}(t, \overline{\Omega}, \overline{\Omega}') := K(t, \cos \chi) - e_0^2 k^{\text{ell}}(t, \overline{\Omega}, \overline{\Omega}') . \qquad (10.40)$$

10.4 The ellipsoidal Poisson kernel

We call the function $K^{\text{ell}}(t, \overline{\Omega}, \overline{\Omega}')$ the ellipsoidal Poisson kernel since it solves Dirichlet's problem (10.1)–(10.3) on an ellipsoid of revolution. The next effort will be devoted to convert the spectral form (10.35) of $k^{\text{ell}}(t, \overline{\Omega}, \overline{\Omega}')$ to a spatial representation.

Substituting eqn.(B.9.21) into eqn.(10.35), the integral kernel $k^{\text{ell}}(t, \overline{\Omega}, \overline{\Omega}')$ can be composed from three different terms:

$$k^{\text{ell}}(t, \overline{\Omega}, \overline{\Omega}') = \frac{1 - t^2}{2} \Big[\sin \beta \left(\cos \beta \sin \chi \cos \chi \cos \alpha - \sin \beta \cos^2 \chi \cos^2 \alpha + \right.$$

$$+ \quad \sin \beta \sin^2 \alpha \Big) L_1(t, \cos \chi) + \left(1 - \sin^2 \beta \sin^2 \alpha \right) L_2(t, \cos \chi) - \qquad (10.41)$$

$$- \quad \sin \beta \cos \alpha \left(\cos \beta \sin \chi - \sin \beta \cos \chi \cos \alpha \right) L_3(t, \cos \chi) \Big] ,$$

where the isotropic parts $L_i(t, \cos \chi)$, $i = 1, 2, 3$, of function $k^{\text{ell}}(t, \overline{\Omega}, \overline{\Omega}')$ are given by infinite series of the Legendre polynomials,

$$L_1(t, x) := \sum_{j=1}^{\infty} \frac{2j - 1}{2j + 1} t^j \frac{dP_j(x)}{dx} , \qquad (10.42)$$

$$L_2(t, x) := \sum_{j=0}^{\infty} \frac{(j + 1)^2 (2j + 1)}{2j + 3} t^{j+1} P_j(x) , \qquad (10.43)$$

$$L_3(t, x) := \sum_{j=1}^{\infty} \frac{j(2j - 1)}{2j + 1} t^j P_j(x) , \qquad (10.44)$$

where x abbreviates the expression:

$$x := \cos \chi = \cos \beta \cos \beta' + \sin \beta \sin \beta' \cos(\lambda - \lambda') . \qquad (10.45)$$

We shall now attempt to sum up infinite series for $L_i(t, x)$. Simple manipulations with expressions (10.42)–(10.44) result in fact that $L_i(t, x)$ can be, besides others, expressed in terms of sums

$$\sum_{j=1}^{\infty} \frac{t^j}{2j + 1} P_j(x) , \qquad \text{and/or} \qquad \sum_{j=1}^{\infty} \frac{t^j}{2j + 1} \frac{dP_j(x)}{dx} . \qquad (10.46)$$

Both the sums are expressible in terms of the incomplete elliptic integrals (Abramowitz and Stegun, 1970, chapt.17) which cannot be evaluated analytically but only numerically by numerical quadrature (Press et al., 1992, sect. 6.11). Hence, our method of summation will follow another idea. Since the kernels $L_i(t, x)$ are singular at the point $(t = 1, x = 1)$, we shall remove the contributions that are responsible for the singular behaviour from sums (10.42)–(10.44). These contributions will be expressed in closed analytical forms. Having removed the singular contributions, the rest will be represented by quickly convergent infinite series, which are bounded on whole interval $-1 \le x \le 1$ and $t = 1$. Prescribing an error of computation, they can simply be summed up numerically.

To sum up the infinite series (10.42)–(10.44) we will use the summation formulae for infinite series of Legendre polynomials and their derivatives listed in Appendix A.10. In order to use them, the fractions occurring in series (10.42)–(10.44) need to be arranged as follows:

$$\frac{2j-1}{2j+1} = 1 - \frac{15}{8j} + \frac{5}{4(j-1)} - \frac{3}{8(j-2)} +$$
$$+ \frac{15}{4(j-2)(j-1)j(2j+1)} \quad (j \ge 3) ,$$

$$\frac{(j+1)^2(2j+1)}{2j+3} = j^2 + j + \frac{1}{2} - \frac{1}{4j} + \frac{3}{4j(2j+3)} , \quad (10.47)$$

$$\frac{j(2j-1)}{2j+1} = j - 1 + \frac{1}{2j} - \frac{1}{2j(2j+1)} .$$

These expressions allows us to write eqns.(10.42)–(10.44) as

$$L_1(t, x) = \frac{\partial S_3(t, x)}{\partial x} - \frac{15}{8} \frac{\partial S_4(t, x)}{\partial x} + \frac{5}{4} \frac{\partial S_5(t, x)}{\partial x} - \frac{3}{8} \frac{\partial S_6(t, x)}{\partial x} + \frac{29}{24} t -$$
$$- \frac{171}{80} t^2 x + R_1(t, x) , \quad (10.48)$$

$$L_2(t, x) = t \left[S_1(t, x) + S_2(t, x) - \frac{1}{2} S_3(t, x) - \frac{1}{4} S_4(t, x) \right] + R_2(t, x) , \quad (10.49)$$

$$L_3(t, x) = S_2(t, x) - S_3(t, x) + \frac{1}{2} S_4(t, x) + R_3(t, x) , \quad (10.50)$$

where residuals $R_i(t, x)$, $i = 1, 2, 3$, are discussed in the next section and sums $S_i(t, x)$, $i = 1, ..., 6$, and their derivatives with respect to variable x are listed in Appendix A.10. Substituting them into eqns.(10.48)–(10.50), we get the final formulae for the integral kernels $L_i(t, x)$, $i = 1, 2, 3$. Instead of writing down these final formulae explicitly, we present subroutine KERL (see Appendix B.10) which computes the kernels $L_i(t, x)$, $i = 1, 2, 3$, and demonstrates the simplicity of their computation. Note that subroutine KERL does not consider the residuals $R_i(t, x)$, $i = 1, 2, 3$, because of their negligible size; they can usually be neglected in practical implementations of the presented theory. However, we show a simple

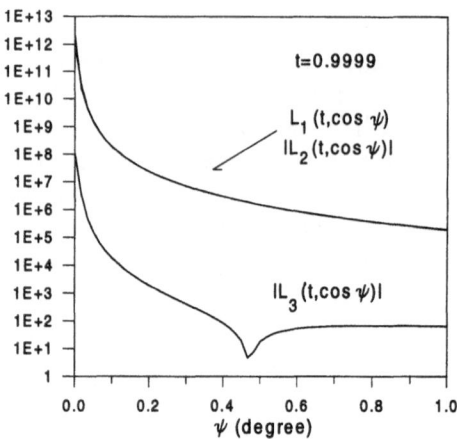

Figure 10.2: The kernels $L_1(t, \cos \chi)$ and $|L_i(t, \cos \chi)|$, $i = 2, 3$, for $t = 0.9999$
and $0 \leq \chi \leq 1°$.

way of computing them in the next section.

Figure 10.2 plots the integral kernel $L_1(t, x)$ and the absolute values of kernels
$L_i(t, x)$, $i = 2, 3$ (in order to use the logarithmic scale on the vertical axis), for
$t = 0.9999$ and $x = \cos \chi$, $0 \leq \chi \leq 1°$. We can observe a huge increase in
amplitudes of $L_i(t, \cos \chi)$ as $\chi \to 0$.

10.5   Residuals $R_i(t, x)$

The residuals $R_i(t, x)$, appearing in eqns.(10.48)–(10.50), are of the forms

$$R_1(t, x) = \sum_{j=3}^{\infty} \frac{15}{4(j-2)(j-1)j(2j+1)} t^j \frac{dP_j(x)}{dx} \,, \tag{10.51}$$

$$R_2(t, x) = \frac{1}{3}t + \sum_{j=1}^{\infty} \frac{3}{4j(2j+3)} t^{j+1} P_j(x) \,, \tag{10.52}$$

$$R_3(t, x) = -\sum_{j=1}^{\infty} \frac{1}{2j(2j+1)} t^j P_j(x) \,. \tag{10.53}$$

Let us show that $R_i(t, x)$ are bounded for $0 < t \leq 1$ and $-1 \leq x \leq 1$. For $-1 \leq
x \leq 1$, the derivatives of Legendre polynomials can be estimated by eqn.(9.66)
which yields

$$|R_1(t, x)| \leq |R_1(1, x)| \leq \frac{15}{4} \sum_{j=3}^{\infty} \frac{j}{(j-2)(j-1)(2j+1)} <$$

$$< \frac{15}{8} \sum_{j=3}^{\infty} \frac{1}{(j-2)(j-1)} = \frac{15}{8} \,, \tag{10.54}$$

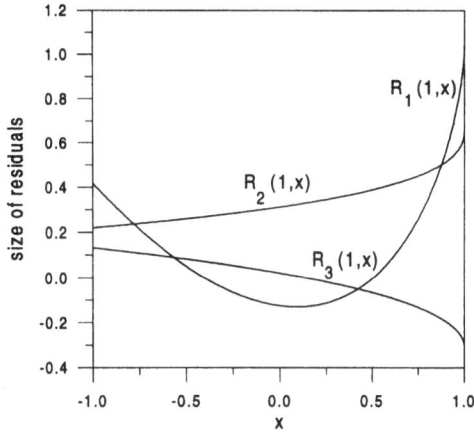

Figure 10.3: The residuals $R_i(1, x)$, $i = 1, 2, 3$, for $x \in \langle -1, 1 \rangle$.

where we have borrowed the summation formula from Mangulis (1965, p.53):

$$\sum_{j=1}^{\infty} \frac{1}{j(j+1)} = 1 . \tag{10.55}$$

Similarly, the size of residual $R_2(t, x)$ can be estimated by realizing that $|P_j(x)| \le 1$:

$$|R_2(t, x)| \le |R_2(1, x)| \le \frac{1}{3} + \frac{3}{4} \sum_{j=1}^{\infty} \frac{1}{j(2j+3)} < \frac{1}{3} + \frac{3}{8} \sum_{j=1}^{\infty} \frac{1}{j^2} = \frac{1}{3} + \frac{3\pi^2}{48} \doteq 0.951 , \tag{10.56}$$

where we have used the summation formula (9.71). An analogous procedure for $R_3(t, x)$ yields:

$$|R_3(t, x)| < \frac{\pi^2}{24} \doteq 0.411 . \tag{10.57}$$

Figure 10.3 plots the residuals $R_i(1, x)$ for $-1 \le x \le 1$. We can see that $|R_1(1, x)| < 1.04$, $|R_2(1, x)| < 0.66$ and $|R_3(1, x)| < 0.31$ which are more precise estimates than those given by eqns.(10.54), (10.56) and (10.57), respectively.

It is important to realize that the the residuals $R_i(t, x)$ are bounded at any point of the domain $(0, 1) \otimes \langle -1, 1 \rangle$ so that the singularity of integral kernels $L_i(t, x)$ at the point $(1, 1)$ are expressed analytically by the first terms on the right-hand side of eqns.(10.48)–(10.50). In addition, the function $R_i(t, x)$ can simply be evaluated numerically, since it is represented by a quickly convergent series. Figure 10.4 demonstrates this fact in a transparent way; it plots the decay of magnitudes of series terms (10.51)–(10.53) at the point $(1, 1)$ with increasing degree j. Inspecting Figure 10.4 we can estimate that it is sufficient to sum up the infinite series (10.51) for $R_1(t, x)$ up to $j \approx 11$ and the series in (10.52) and (10.53) for $R_2(t, x)$ and for $R_3(t, x)$ up to $j \approx 3$ in order to achieve an absolute accuracy

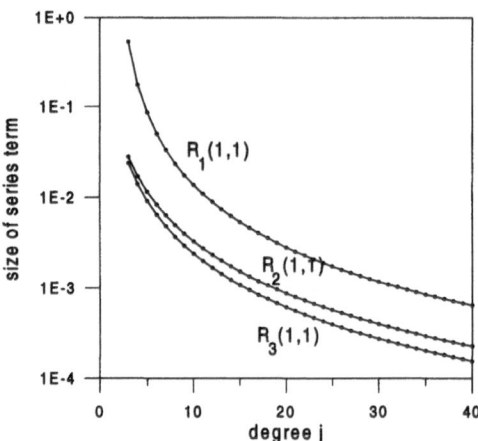

Figure 10.4: The sizes of particular terms creating the infinite series for
$R_i(t, x)$, $i = 1, 2, 3$, at the point $(1, 1)$.

of the order of 0.01. This accuracy is sufficient for evaluating the ellipsoidal
Poisson kernel $K^{\text{ell}}(t, \overline{\Omega}, \overline{\Omega}')$ in the frame of the $O(e_0^2)$-approximation.

10.6   The behaviour at the singularity

Though the spherical Poisson integral as well as the ellipsoidal Poisson integral
(10.39) are regular, which means that there is no singularity of the integral kernel
inside the integration domain, it may simply happen that the distance of the
computation point from the reference sphere or ellipsoid becomes small. Then
the singularity of the integral kernels lie in a close neighbourhood of the integration domain and the integrals turn to be *nearly singular* (Hackbusch, 1995,
sect.9.4.3). As a matter of fact, the integration error of a quadrature method for
computing nearly singular integrals depends, besides other factors, on the degree
of singularity of a integral kernel. Hence, we may rise the question, whether the
degree of singularity of the original spherical Poisson kernel is the same as the
degree of singularity of the ellipsoidal Poisson kernel. The answer to this question
gives us a hint as to whether we can apply a numerical quadrature method of
the same precision for computing both the spherical and the ellipsoidal Poisson
integral.

The singularity of the Poisson kernels $K(t, \cos \chi)$ and $K^{\text{ell}}(t, \overline{\Omega}, \overline{\Omega}')$ lies at the
point $(t = 1, \chi = 0)$, or, equivalently, $(t = 1, \overline{\Omega}' = \overline{\Omega})$. To determine the degree of
singularity of $K^{\text{ell}}(t, \overline{\Omega}, \overline{\Omega}')$, let us study the behaviour of the kernels $L_i(t, \cos \chi)$
at the singular point. Using formulae (10.48)–(10.50), we can, after some algebra,
derive that

$$L_i(t, \cos \chi)\Big|_{\substack{t \to 1 \\ \chi \to 0}} \quad \sim \quad \frac{1}{d^3} \qquad \text{for} \quad i = 1, 2 , \qquad (10.58)$$

$$L_3(t, \cos\chi)|_{\substack{t \to 1 \\ \chi \to 0}} \sim \frac{1}{d^2}, \tag{10.59}$$

where

$$d = \sqrt{(1-t)^2 + \chi^2}. \tag{10.60}$$

Consequently, the kernel $k^{\text{ell}}(t, \overline{\Omega}, \overline{\Omega}')$ behaves as follows:

$$k^{\text{ell}}(t, \overline{\Omega}, \overline{\Omega}')\Big|_{\substack{t \to 1 \\ \overline{\Omega}' \to \overline{\Omega}}} \sim \frac{1 - t^2}{d^3}. \tag{10.61}$$

Equation (10.33) shows that

$$K(t, \cos\chi)\Big|_{\substack{t \to 1 \\ \overline{\Omega}' \to \overline{\Omega}}} \sim \frac{1 - t^2}{d^3}. \tag{10.62}$$

We can conclude that the ellipsoidal Poisson kernel $K^{\text{ell}}(t, \overline{\Omega}, \overline{\Omega}')$ has the same degree of singularity at the point $(t = 1, \overline{\Omega}' = \overline{\Omega})$ as the original spherical Poisson kernel at the same point; the estimate (10.36) confirms this conclusion.

10.7 Conclusion

The chapter was motivated by the question as to whether Green's function to the external Dirichlet boundary-value problem for the Laplace equation with data distributed on an ellipsoid of revolution can be constructed. To answer this question, we first constructed the solution to this boundary-value problem in terms of ellipsoidal harmonics. The fact that the solution is represented by slowly convergent series of ellipsoidal harmonics prevents to use it for regional geoid computation. That is why, in the next step, we have confined ourselves to the $O(e_0^2)$-approximation, meaning that we kept terms of the magnitudes of the order of e_0^2; the terms of the order of $O(e_0^4)$ and of higher powers have not been considered. Nevertheless, the accuracy of the order of $O(e_0^2)$ is fairly good for today's requirement on geoid height computation. Within this accuracy, we have shown that the solution can be written as an integral taken over the full solid angle from data distributed on an ellipsoid of revolution which are multiplied by a kernel consisting of the traditional spherical Poisson kernel and the correction due to the elliptical shape of the boundary; we call the whole resolving kernel as the ellipsoidal Poisson kernel. We have managed to express the ellipsoidal Poisson kernel, originally represented in the form of an infinite sum of ellipsoidal harmonics, as a finite combination of elementary functions analytically describing the behaviour of the ellipsoidal Poisson kernel at the singular point. Such an expression is suitable for numerical solution to the harmonic upward and downward continuation of gravitation. Moreover, we have shown that the degree of singularity of the ellipsoidal Poisson kernel in the vicinity of its singular point is of the same degree as that of the original spherical Poisson kernel.

A.10   Some sums of infinite series of Legendre polynomials

In this appendix, let us recall some summation formulae for infinite series of the Legendre polynomials (other can be found e.g. in Pick et al. (1973, sect.D-18), or in Thong (1993)):

$$S_1(t,x) := \sum_{j=1}^{\infty} j^2\, t^j P_j(x) \;=\; \frac{t(t+x)}{g^3} - \frac{3t^2(1-x^2)}{g^5}\,, \qquad (A.10.1)$$

$$S_2(t,x) := \sum_{j=1}^{\infty} j\, t^j P_j(x) \;=\; -\frac{t(t-x)}{g^3}\,, \qquad (A.10.2)$$

$$S_3(t,x) := \sum_{j=1}^{\infty} t^j P_j(x) \;=\; \frac{1}{g} - 1\,, \qquad (A.10.3)$$

$$S_4(t,x) := \sum_{j=1}^{\infty} \frac{t^j}{j} P_j(x) \;=\; -\ln\left(\frac{g+1-tx}{2}\right)\,, \qquad (A.10.4)$$

$$S_5(t,x) := \sum_{j=2}^{\infty} \frac{t^j}{j-1} P_j(x) \;=\; 1 - g - tx - tx \ln\left(\frac{g+1-tx}{2}\right)\,, \qquad (A.10.5)$$

$$S_6(t,x) := \sum_{j=3}^{\infty} \frac{t^j}{j-2} P_j(x) \;=\; \frac{1}{2} + tx - \frac{7t^2x^2}{4} + \frac{t^2}{4} - \frac{1}{2}(1+3tx)g -$$
$$- \frac{t^2}{2}\left(3x^2 - 1\right)\ln\left(\frac{g+1-tx}{2}\right)\,, \qquad (A.10.6)$$

where $g \equiv g(t,x) := \sqrt{1+t^2-2tx}$, $-1 \le x \le 1$, and $0 < t \le 1$. The derivatives of the sums S_i, $i = 2, ..., 6$, with respect to variable x read

$$\frac{\partial S_2(t,x)}{\partial x} = \frac{t}{g^3} - \frac{3t^2(t-x)}{g^5}\,, \qquad (A.10.7)$$

$$\frac{\partial S_3(t,x)}{\partial x} = \frac{t}{g^3}\,, \qquad (A.10.8)$$

$$\frac{\partial S_4(t,x)}{\partial x} = \frac{t(g+1)}{g(g+1-tx)}\,, \qquad (A.10.9)$$

$$\frac{\partial S_5(t,x)}{\partial x} = \frac{t}{g} - t + \frac{t^2x(g+1)}{g(g+1-tx)} - t\ln\left(\frac{g+1-tx}{2}\right)\,, \qquad (A.10.10)$$

$$\frac{\partial S_6(t,x)}{\partial x} = t - \frac{7t^2x}{2} + \frac{t}{2g}(1+3tx) - \frac{3tg}{2} + \frac{t^3\left(3x^2-1\right)(g+1)}{2g(g+1-tx)} -$$
$$- 3t^2x\ln\left(\frac{g+1-tx}{2}\right)\,. \qquad (A.10.11)$$

B.10   Program KERL

```
c————————————————————————————————————————————
      subroutine KERL(t,x,ql)
c
c     Given values t and x, (0 < t ≤ 1, −1 ≤ x ≤ 1),
c     this subroutine returns the integration kernels L_i(t,x),
c     i = 1, 2, 3, evaluated at (t, x) as ql(i), i=1,2,3.
c————————————————————————————————————————————
      implicit real*8 (a-h,o-z)
      dimension ql(3)
        g=dsqrt(1d0+t*t-2d0*t*x)
        w=g+1d0-t*x
      sum1=t*(t+x)/(g**3)-3d0*t*t*(1d0-x*x)/(g**5)
      sum2=-t*(t-x)/(g**3)
      sum3=(1d0/g-1d0)
      sum4=-dlog(w/2d0)
      dsum3=t/(g**3)
      dsum4=t*(g+1d0)/g/w
      dsum5=t/g-t+t*t*x*(g+1d0)/g/w+t*sum4
      dsum6=t-7d0*t*t*x/2d0+t*(1d0+3d0*t*x)/2d0/g-3d0*t*g/2d0
     *   +(3d0*x*x-1d0)*t*t*t*(g+1d0)/g/w/2d0+3d0*t*t*x*sum4
      ql(1)=dsum3-15d0*dsum4/8d0+5d0*dsum5/4d0-3d0*dsum6/8d0
     *   +29d0*t/24d0-171d0*t*t*x/80d0
      ql(2)=t*(sum1+sum2-sum3/2d0-sum4/4d0)
      ql(3)=sum2-sum3+sum4/2d0
      return
      end
```

Chapter 11

The Stokes boundary-value problem with ellipsoidal corrections in boundary condition

Prescribing the surface gravity potential, the magnitude of surface gravity, the volumetric density of topographical masses and the angular velocity of the Earth's rotation, the scalar, free boundary-value problem for determination of the geoid can be formulated. This problem, sometimes called the problem for gravimetric determination of the geoid, is non-linear with respect to sought gravity potential since the magnitude of gravity is a non-linear function of generating potential. The problem may be linearized with respect to a reference gravity GM/r^2, which is quoted as a spherical approximation in gravity space. The error of this linearization may reach 5 mgal counted in gravity. To reduce the linearization error of spherical approximation to a level of the order of 0.2 mgal, the gravity field of a level ellipsoid, the so-called Somigliana-Pizzetti normal gravity, is to be introduced as a reference. Compared to the spherical approximation, the linearized boundary condition for anomalous gravity contains two additional terms, the so-called ellipsoidal corrections (e.g., Jekeli, 1981), the magnitudes of which are of the order of the squared eccentricity of the level ellipsoid.

Having linearized the boundary-value problem for gravimetric determination of the geoid in gravity space, the problem is still non-linear (and also free) in geometry space since the geoid to be determined is a non-linear function of a sought anomalous gravitational potential. Hence, an additional linearization, this time in geometry space, is convenient to introduce. Only two geometrical approximations of the geoid in the boundary condition for anomalous gravity have been introduced so far, namely, the approximation of the geoid by a sphere or by an ellipsoid of revolution. Both linearize (and also fix) the problem with respect to the geometrical shape of the geoid; they may thus be termed as the spherical or ellipsoidal approximation in geometry space. The spherical approximation of the geoid produces the relative error of the order of 3×10^{-3} which may cause an absolute error reaching 0.5 m counted in geoidal heights. Approximating the figure of the geoid in the boundary condition for the anomalous gravity by an ellipsoid of revolution reduces the linearization error in geometry space since the geoid deviates from an ellipsoid of revolution much less than from a sphere. The ellipsoidal linearization may cause relative errors of the order of 1.5×10^{-5}; the absolute error in geoidal heights does not exceed 2 mm.

Taking into account different types of linearization in gravity and geometry space, the originally non-linear, free boundary-value problem for determination of gravimetric geoid may be approximated by various types of linearized, fixed boundary-value problems. Adopting the spherical approximation in both gravity and geometry spaces results in the well-known spherical Stokes problem, firstly formulated and solved by Stokes (1849).

The linearization process of the boundary-value problem for gravimetric geoid determination may follow another possibility: the ellipsoidal approximation in gravity space and the spherical approximation in geometry space. This leads to a boundary-value problem of Stokes's type with two ellipsoidal corrections involved in boundary condition. These additional terms couple the spherical harmonics of degree two with spectral content of sought anomalous gravitational potential. Due to this coupling effect, there is no exact analytical solution to the Stokes boundary-value problem when the ellipsoidal corrections are involved. An iterative solution to this problem is usually recommended in geodetic literature (e.g., Cruz, 1985; Heck, 1991). At the initial step, the ellipsoidal correction terms are computed from apriorily known potential, e.g., global gravity model, and subtracted from the right-hand side of boundary condition for anomalous gravity. Then the solution is looked for by successive iterations evaluating the ellipsoidal corrections from the preceeding iterative step. However, the convergency of this iterative scheme has not been proved.

Instead of applying the iterative solution, in this chapter (see also Martinec 1998) we shall attempt to construct Green's function associated with the problem. As discussed above, an exact Green's function cannot be found due to the coupling effect between spherical harmonics of degree two and spherical harmonics of anomalous gravitational potential. That is why, Green's function will be constructed approximately retaining the terms of magnitudes up the order of Earth's flattening. Neglecting higher-order terms causes relative errors of the order of 1.5×10^{-5}; the absolute error in geoidal heights does not exceed 2 mm.

Two remarks should be noted in the context. The construction of Green's function allows us to avoid to use an iterative approach in solving Stokes's boundary-value problem with ellipsoidal corrections involved in boundary condition as proposed by Cruz (1985) or Heck (1991). The iterative scheme is questionable from numerical point of view since the convergency of iterations has not been proved yet. In Chapter 1 we demonstrate that the iterative scheme fails when it is applied to solve the Stokes two-boundary value problem that arises when the heights of the Earth's topography enter the boundary-value problem for gravimetric geoid determination instead of the geopotential numbers. In this case, the ellipsoidal correction terms cannot be computed neither from the previous iterative step nor from a known global gravitational model of the Earth.

There is even another possibility of how to linearize the originally non-linear, free boundary-value problem, namely, to use the ellipsoidal approximation in both gravity and geometry spaces. This approach seems to be the most precise compared to the other ways of linearization discussed above since the linearization errors are equal to 0.2 mgal and 2 mm in gravity and geometry spaces, respectively. However, this approximation does not remove the ellipsoidal correction terms in the boundary condition for anomalous gravity. They have different analytical forms compared to the formulae introduced below in eqn.(11.2) and depend on the type of the projection of the Earth's surface onto the geoid; their analytical forms are derived in Heck (1991).

In Chapter 9 we made a first step in solving this type of boundary-value problem. We have not considered the ellipsoidal correction terms in boundary condition for anomalous gravity, which corresponds to linearization in gravity space by the reference gravity GM/u^2 (u is the first ellipsoidal coordinate), and constructed Green's function to this problem within the accuracy of the order of $O(e_0^2)$. We hope to report about the progress in solving the ellipsoidal Stokes problem when the ellipsoidal corrections are also included in boundary condition for anomalous gravity in near future.

11.1 Formulation of the boundary-value problem

We shall solve the boundary-value problem of Stokes's type with the ellipsoidal corrections involved in boundary condition in the geocentric spherical coordinates (r, Ω), where Ω stands for the pair of angular coordinates, co-latitude ϑ and longitude λ. The potential $T(r, \Omega)$ to be determined <u>on and outside</u> the reference sphere $r = R$ is governed by the following boundary-value problem:

$$\nabla^2 T = 0 \qquad \text{for } r > R , \qquad (11.1)$$

$$\frac{\partial T}{\partial r} + \frac{2}{r} T - \varepsilon_h(T) - \varepsilon_\gamma(T) = -f \qquad \text{for } r = R , \qquad (11.2)$$

$$T = \frac{c}{r} + O\left(\frac{1}{r^3}\right) \qquad \text{for } r \to \infty , \qquad (11.3)$$

where the ellipsoidal correction $\varepsilon_h(T)$, defined by eqn.(1.30), occurs due to difference between the derivative of the potential T with respect to the plumbline of the normal gravity field and the radial derivative of T, whereas the ellipsoidal correction $\varepsilon_\gamma(T)$, defined by eqn.(1.40), arises due to the upward continuation of the normal gravity field from the reference ellipsoid of revolution to the Earth's surface. These terms have been introduced by a couple of authors to approximate the originally non-linear boundary-value problem for geoid determination by means of the linearized problem (11.1)–(11.3) with the accuracy of the order of Earth's eccentricity. The reader is referred to Moritz (1980a), Jekeli (1981) or Heck (1991) for more details.

We assume that $f(\Omega)$ is a known square-integrable function, i.e., $f(\Omega) \in L_2(\Omega)$, which can be obtained from measurements of the gravity field on the Earth's surface reduced by the attraction of the reference ellipsoid of revolution and the topographical masses. The first-degree harmonics of T have been removed from the solution in order to guarantee the uniqueness of the solution (c in eqn.(11.3) is a constant). Furthermore, we can easily deduce that the solution to the problem (11.1)–(11.3) exists only if the first-degree spherical harmonics of $f(\Omega)$ are removed by the postulate

$$\int_{\Omega_0} f(\Omega)Y_{1m}^*(\Omega)d\Omega = 0 \qquad \text{for } m = -1,0,1 , \tag{11.4}$$

where $Y_{1m}(\Omega)$ are spherical harmonics of the first-degree and order m, Ω_0 is the full solid angle, $d\Omega = \sin\vartheta d\vartheta d\lambda$, and asterisk stands for complex conjugation. Throughout this chapter we will assume that conditions (11.4) are satisfied.

The solution to the Laplace equation (11.1) can be written in terms of solid spherical harmonics $r^{-j-1}Y_{jm}(\Omega)$ as follows

$$T(r,\Omega) = \sum_{\substack{j=0 \\ j \neq 1}}^{\infty} \left(\frac{R}{r}\right)^{j+1} \sum_{m=-j}^{j} T_{jm}Y_{jm}(\Omega) , \tag{11.5}$$

where T_{jm} are expansion coefficients to be determined from the boundary condition (11.2). To satisfy the asymptotic condition (11.3), the term with degree $j = 1$ is excluded from summation over j's.

Substituting expansion (11.5) and eqn.(A.11.8), derived in Appendix A.11, for harmonic representation of the ellipsoidal corrections into the boundary condition (11.2), we get

$$\frac{1}{R} \sum_{\substack{j=0 \\ j \neq 1}}^{\infty} (j-1) \sum_{m=-j}^{j} T_{jm}Y_{jm}(\Omega)+$$

$$+\frac{e_0^2}{R} \sum_{\substack{j=0 \\ j \neq 1}}^{\infty} \sum_{m=-j}^{j} \left\{ \frac{j+1}{2j-1}\sqrt{\frac{[(j-1)^2-m^2][j^2-m^2]}{(2j-3)(2j+1)}} \, T_{j-2,m}+ \right.$$

$$+ \left[\frac{j(j+1)-3m^2}{(2j-1)(2j+3)}-1\right] T_{jm}- \tag{11.6}$$

$$\left. -\frac{j}{2j+3}\sqrt{\frac{[(j+1)^2-m^2][(j+2)^2-m^2]}{(2j+1)(2j+5)}} \, T_{j+2,m} \right\} Y_{jm}(\Omega) = -f(\Omega) .$$

Moreover, expanding function $f(\Omega)$ in a series of spherical harmonics,

$$f(\Omega) = \sum_{j=0}^{\infty} \sum_{m=-j}^{j} \int_{\Omega_0} f(\Omega')Y_{jm}^*(\Omega')d\Omega'Y_{jm}(\Omega) , \tag{11.7}$$

where the first-degree spherical harmonics of $f(\Omega)$ are equal to zero due to the assumption (11.4), substituting expansion (11.7) into eqn.(11.6), and comparing the coefficients at spherical harmonics $Y_{jm}(\Omega)$ we end up with the infinite system of linear algebraic equations for coefficients T_{jm}:

$$e_0^2 a_{jm} T_{j-2,m} + (1 + e_0^2 b_{jm}) T_{jm} + e_0^2 c_{jm} T_{j+2,m} = d_{jm} , \qquad (11.8)$$

$j = 0, 2, ..., m = -j, -j + 1, ..., j$, where system matrix elements are equal to

$$a_{jm} := \frac{j+1}{(j-1)(2j-1)} \sqrt{\frac{[(j-1)^2 - m^2][j^2 - m^2]}{(2j-3)(2j+1)}} \qquad (11.9)$$

$$b_{jm} := \frac{3}{j-1} \left[\frac{(j+1)^2 - m^2}{(2j-1)(2j+3)} - \frac{j}{2j-1} \right] , \qquad (11.10)$$

$$c_{jm} := -\frac{j}{(j-1)(2j+3)} \sqrt{\frac{[(j+1)^2 - m^2][(j+2)^2 - m^2]}{(2j+1)(2j+5)}} , \qquad (11.11)$$

and the right-hand side are given by surface integrals

$$d_{jm} := \frac{R}{j-1} \int_{\Omega_0} f(\Omega') Y_{jm}^*(\Omega') d\Omega'. \qquad (11.12)$$

11.2   The $O(e_0^2)$-approximation

From practical point of view, the spectral form (11.5) of the solution to the boundary-value problem (11.1)–(11.3) is often inconvenient, since the construction of the spectral components of $f(\Omega)$ and solving the system of equations (11.8) for some cut-off degree $j = j_{max}$ may become time consuming. Moreover, in the case that the magnitudes of ellipsoidal correction terms $\varepsilon_h(T)$ and $\varepsilon_\gamma(T)$ in boundary condition (11.2) are much smaller then the Stokes term $\partial T/\partial r + 2T/r$, which is the case of the Earth, the solution to our problem should be close to the solution to the spherical Stokes problem. We will thus attempt to rewrite $T(r, \Omega)$ as a sum of the well-known Stokes integral plus the contribution due to the corrections $\varepsilon_h(T)$ and $\varepsilon_\gamma(T)$. An evident advantage of such a decomposition is that existing theories as well as numerical codes for geoid height computation can simply be corrected for the ellipsoidal correction terms.

To build up the theory as simple as possible but still matching the requirements on geoid height accuracy, throughout following derivations we will keep the terms of the magnitudes of the order of $O(e_0^2)$ and neglect terms of higher powers of e_0^2. This approximation is justifiable because the errors introduced by this approximation are at most 1.5×10^{-5} which then causes an error of at most a few millimeters in the geoidal heights.

Inspecting eqns.(11.8) we can see that this system is not coupled with respect to the order m and can thus be solved separately for a individual order m. Furthermore, the equations are also separated for even and odd j's, and again can be solved separately for even and odd angular degrees j. In matrix notation, the set of equations (11.8) has a tridiagonal form:

$$
\begin{pmatrix}
1 + e_0^2 b_{mm} & e_0^2 c_{mm} & 0 & \cdots \\
e_0^2 a_{m+2,m} & 1 + e_0^2 b_{m+2,m} & e_0^2 c_{m+2,m} & \cdots \\
0 & e_0^2 a_{m+4,m} & 1 + e_0^2 b_{m+4,m} & e_0^2 c_{m+4,m} & \cdots \\
& & \cdots
\end{pmatrix} \times
$$

$$
\times
\begin{pmatrix}
T_{mm} \\
T_{m+2,m} \\
T_{m+4,m} \\
\vdots
\end{pmatrix}
=
\begin{pmatrix}
d_{mm} \\
d_{m+2,m} \\
d_{m+4,m} \\
\vdots
\end{pmatrix},
\tag{11.13}
$$

$m = 0, 2, 3, \dots$. A simple analysis of eqns.(11.9)–(11.11) reveals that the magnitudes of matrix elements can be estimated as $(j \neq 1)$

$$
|a_{jm}| < 1, \qquad |b_{jm}| \leq 1, \qquad |c_{jm}| < 1.
\tag{11.14}
$$

Assuming that $e_0^2 \ll 1$, the sizes of the off-diagonal elements in the system (11.13) are at least $1/e_0^2$-times smaller than those of the diagonal elements. As a consequence, the solution to the system of equations (11.13) always exists and is unique provided, again, that $j \neq 1$ and $e_0^2 \ll 1$.

A special structure of matrix of system (11.13) allows us to solve the system approximately such that the accuracy of the order of e_0^2 is maintained in the solution. In Appendix B.11, we present such an approach resulting in eqn.(B.11.10). Applying this result to system (11.13), the solution is

$$
T_{jm} = d_{jm} - e_0^2 (b_{jm} d_{jm} + a_{jm} d_{j-2,m} + c_{jm} d_{j+2,m}) + O(e_0^4).
\tag{11.15}
$$

Inserting coefficients T_{jm} into eqn.(11.5) yields

$$
T'(r, \Omega) = \sum_{\substack{j=0 \\ j \neq 1}}^{\infty} \left(\frac{R}{r}\right)^{j+1} \sum_{m=-j}^{j} d_{jm} Y_{jm}(\Omega) -
$$

$$
- e_0^2 \sum_{\substack{j=0 \\ j \neq 1}}^{\infty} \left(\frac{R}{r}\right)^{j+1} \sum_{m=-j}^{j} (b_{jm} d_{jm} + c_{jm} d_{j+2,m}) Y_{jm}(\Omega) -
\tag{11.16}
$$

$$
- e_0^2 \sum_{\substack{j=0 \\ j \neq 1}}^{\infty} \left(\frac{R}{r}\right)^{j+3} \sum_{m=-j}^{j} a_{j+2,m} d_{jm} Y_{j+2,m}(\Omega) + O(e_0^4).
$$

Substituting for coefficients d_{jm} from eqn.(11.12), interchanging the order of summation over j and m with integration over Ω' due to the uniform convergence of the series (11.16), the solution to the boundary-value problem (11.1)–(11.3) within the accuracy of the order of $O(e_0^2)$ reads

$$
T(r, \Omega) = \frac{R}{4\pi} \int_{\Omega_0} f(\Omega') \Bigg[4\pi \sum_{\substack{j=0 \\ j \neq 1}}^{\infty} \frac{1}{j-1} \left(\frac{R}{r}\right)^{j+1} \sum_{m=-j}^{j} Y_{jm}(\Omega) Y_{jm}^*(\Omega') -
$$

$$-e_0^2 \left(S_{00}^{\text{elco}}(r,\Omega,R,\Omega') + S^{\text{elco}}(r,\Omega,R,\Omega') \right) \Big] d\Omega' , \qquad (11.17)$$

where

$$S_{00}^{\text{elco}}(r,\Omega,R,\Omega') := 4\pi \left[-\frac{R}{r} b_{00} Y_{00}(\Omega) Y_{00}^*(\Omega') + \frac{R}{r} c_{00} Y_{00}(\Omega) Y_{20}^*(\Omega') - \right.$$
$$\left. - \left(\frac{R}{r}\right)^3 a_{20} Y_{20}(\Omega) Y_{00}^*(\Omega') \right] , \qquad (11.18)$$

and

$$S^{\text{elco}}(r,\Omega,R,\Omega') := 4\pi \sum_{j=2}^{\infty} \left(\frac{R}{r}\right)^{j+1} \sum_{m=-j}^{j} \left[\frac{b_{jm}}{j-1} Y_{jm}(\Omega) Y_{jm}^*(\Omega') + \right.$$

$$\left. + \frac{c_{jm}}{j+1} Y_{jm}(\Omega) Y_{j+2,m}^*(\Omega') + \left(\frac{R}{r}\right)^2 \frac{a_{j+2,m}}{j-1} Y_{j+2,m}(\Omega) Y_{jm}^*(\Omega') \right] . \qquad (11.19)$$

Using the Laplace addition theorem for spherical harmonics (e.g., Varshalovich et al., 1989, sect.5.17.2),

$$P_j(\cos\psi) = \frac{4\pi}{2j+1} \sum_{m=-j}^{j} Y_{jm}(\Omega) Y_{jm}^*(\Omega') , \qquad (11.20)$$

where $P_j(\cos\psi)$ is the Legendre polynomial of degree j, and ψ is the angular distance between directions Ω and Ω', the first term in the square brackets in eqn.(11.17) is equal to the *inhomogeneous spherical Stokes's function* $S(r,\psi,R)$ (Moritz, 1980a, sect.44),

$$S(r,\psi,R) := \sum_{\substack{j=0 \\ j\neq 1}}^{\infty} \frac{2j+1}{j-1} \left(\frac{R}{r}\right)^{j+1} P_j(\cos\psi) = \qquad (11.21)$$

$$= \frac{2R}{\ell} - \frac{3R^2}{r^2} \cos\psi \ln\left(\frac{\ell+r-R\cos\psi}{2r}\right) - \frac{3R\ell}{r^2} - \frac{5R^2}{r^2}\cos\psi , \qquad (11.22)$$

where

$$\ell := \sqrt{r^2 + R^2 - 2rR\cos\psi} . \qquad (11.23)$$

Furthermore, substituting for $b_{00} = 1$, $c_{00} = 0$, $a_{20} = \sqrt{4/5}$, $Y_{00}(\Omega) = 1/\sqrt{4\pi}$, and $Y_{20}(\Omega) = \sqrt{5/\pi}(3\cos^2\vartheta - 1)/4$ into eqn.(11.18), the kernel $S_{00}^{\text{elco}}(r,\Omega,R,\Omega')$ becomes

$$S_{00}^{\text{elco}}(r,\Omega,R,\Omega') = -\frac{R}{r} - \left(\frac{R}{r}\right)^3 (3\cos^2\vartheta - 1) . \qquad (11.24)$$

Finally, the solution (11.17) to the boundary-value problem (11.1)–(11.3) can be written as

$$
\begin{aligned}
T(r,\Omega) \; &= \; \frac{R}{4\pi}\int_{\Omega_0} f(\Omega')S(r,\psi,R)d\Omega' + \\
&+ \; e_0^2\left[\frac{R}{r} + \left(\frac{R}{r}\right)^3(3\cos^2\vartheta - 1)\right]\frac{R}{4\pi}\int_{\Omega_0} f(\Omega')d\Omega' - \\
&- \; e_0^2\frac{R}{4\pi}\int_{\Omega_0} f(\Omega')S^{\text{elco}}(r,\Omega,R,\Omega')d\Omega' \; .
\end{aligned}
\tag{11.25}
$$

In particular, we are interested in finding the potential $T(r,\Omega)$ on the reference sphere $r = R$, i.e., function $T(R,\Omega)$. In this case, the general formula (11.25) reduces to

$$
T(R,\Omega) = \frac{R}{4\pi}\int_{\Omega_0} f(\Omega')S(\psi)d\Omega' +
$$

$$
+3e_0^2\cos^2\vartheta\,\frac{R}{4\pi}\int_{\Omega_0} f(\Omega')d\Omega' - e_0^2\frac{R}{4\pi}\int_{\Omega_0} f(\Omega')S^{\text{elco}}(\Omega,\Omega')d\Omega' \; ,
\tag{11.26}
$$

where

$$
S(\psi) := S(R,\psi,R) = \frac{1}{\sin\frac{\psi}{2}} - 3\cos\psi\ln\left(\sin\frac{\psi}{2} + \sin^2\frac{\psi}{2}\right) - 6\sin\frac{\psi}{2} - 5\cos\psi \; ,
$$

$$
\tag{11.27}
$$

is the *homogeneous spherical Stokes function* (Heiskanen and Moritz, 1967, eqn.(2-164)), and

$$
S^{\text{elco}}(\Omega,\Omega') := S^{\text{elco}}(R,\Omega,R,\Omega') =
$$

$$
= 4\pi\sum_{j=2}^{\infty}\sum_{m=-j}^{j}\left[\frac{b_{jm}}{j-1}Y_{jm}(\Omega)Y_{jm}^*(\Omega') + \frac{c_{jm}}{j+1}Y_{jm}(\Omega)Y_{j+2,m}^*(\Omega') + \right.
\tag{11.28}
$$

$$
\left. +\frac{a_{j+2,m}}{j-1}Y_{j+2,m}(\Omega)Y_{jm}^*(\Omega')\right] \; .
$$

11.3 The 'spherical-ellipsoidal' Stokes function

We shall call the function $S^{\text{elco}}(\Omega,\Omega')$ the 'spherical-ellipsoidal' Stokes function because it describes the effect of the ellipsoidal correction terms $\varepsilon_h(T)$ and $\varepsilon_\gamma(T)$ on the solution to spherical Stokes's boundary-value problem. The next effort will be devoted to convert the spectral form (11.28) of $S^{\text{elco}}(\Omega,\Omega')$ to a spatial representation.

Substituting for a_{jm}, b_{jm}, and c_{jm} from eqns.(11.9)–(11.11) into eqn.(11.28), we have

$$
S^{\text{elco}}(\Omega,\Omega') = 4\pi\sum_{j=2}^{\infty}\sum_{m=-j}^{j}\left\{\frac{3}{(j-1)^2}\left[\frac{(j+1)^2 - m^2}{(2j-1)(2j+3)} - \frac{j}{2j-1}\right]Y_{jm}(\Omega)Y_{jm}^*(\Omega') - \right.
$$

$$-\frac{j}{(j^2-1)(2j+3)}\sqrt{\frac{[(j+1)^2-m^2][(j+2)^2-m^2]}{(2j+1)(2j+5)}}Y_{jm}(\Omega)Y_{j+2,m}^*(\Omega')+ \quad (11.29)$$

$$+\frac{j+3}{(j^2-1)(2j+3)}\sqrt{\frac{[(j+1)^2-m^2][(j+2)^2-m^2]}{(2j+1)(2j+5)}}Y_{j+2,m}(\Omega)Y_{jm}^*(\Omega')\bigg\} \ .$$

One of the trickiest step to find the spatial representation of $S^{elco}(\Omega,\Omega')$ is to sum up the series over m occurring in eqn.(11.29). Making use of the Laplace addition theorem for spherical harmonics, we present the approach for summing these series in Appendix C.11. Substituting from eqns.(C.11.22)–(C.11.24) into eqn.(11.29), the 'spherical-ellipsoidal' Stokes function can be composed from seven different terms

$$S^{elco}(\Omega,\Omega') = \sum_{i=1}^{7} h_i(\vartheta,\psi,\alpha) M_i(\cos\psi) , \qquad (11.30)$$

where ψ and α are polar coordinates, and

$$h_1(\vartheta,\psi,\alpha) := \sin^2\vartheta\left(\cos^2\alpha-\sin^2\alpha\right) , \qquad (11.31)$$

$$h_2(\vartheta,\psi,\alpha) := \cos^2\vartheta\sin\psi - 2\sin\vartheta\cos\vartheta\cos\psi\cos\alpha -$$
$$- \sin^2\vartheta\sin\psi\cos^2\alpha , \qquad (11.32)$$

$$h_3(\vartheta,\psi,\alpha) := \cos^2\vartheta\cos\psi + 2\sin\vartheta\cos\vartheta\sin\psi\cos\alpha -$$
$$- \sin^2\vartheta\cos\psi\cos^2\alpha , \qquad (11.33)$$

$$h_4(\vartheta,\psi,\alpha) := \sin\vartheta\left(\cos\vartheta\sin\psi\cos\psi\cos\alpha - \sin\vartheta\cos^2\psi\cos^2\alpha+\right.$$
$$\left. + \sin\vartheta\sin^2\alpha\right) , \qquad (11.34)$$

$$h_5(\vartheta,\psi,\alpha) := -\sin\vartheta\cos\alpha\left(\cos\vartheta\sin\psi-\sin\vartheta\cos\psi\cos\alpha\right) , \qquad (11.35)$$

$$h_6(\vartheta,\psi,\alpha) := 1-\sin^2\vartheta\sin^2\alpha , \qquad (11.36)$$

$$h_7(\vartheta,\psi,\alpha) := -1 . \qquad (11.37)$$

The isotropic parts $M_i(\cos\psi)$, $i=1,...,7$, of $S^{elco}(\Omega,\Omega')$ are given as infinite series of the Legendre polynomials and their derivatives:

$$M_1(\cos\psi) := \sum_{j=2}^{\infty}\frac{3}{(j^2-1)(2j+3)}\frac{dP_{j+1}(\cos\psi)}{d\cos\psi} , \qquad (11.38)$$

$$M_2(\cos\psi) := \sin\psi\sum_{j=2}^{\infty}\frac{2(j^2+3j+3)}{(j^2-1)(2j+3)}\frac{dP_{j+1}(\cos\psi)}{d\cos\psi} , \qquad (11.39)$$

$$M_3(\cos\psi) := \sum_{j=2}^{\infty}\frac{3(j+2)}{(j-1)(2j+3)}P_{j+1}(\cos\psi) , \qquad (11.40)$$

$$M_4(\cos\psi) := \sum_{j=2}^{\infty}\frac{3(2j+1)}{(j-1)^2(2j-1)(2j+3)}\frac{dP_{j+1}(\cos\psi)}{d\cos\psi} , \qquad (11.41)$$

$$M_5(\cos\psi) := \sum_{j=2}^{\infty}\frac{3(j+1)(2j+1)}{(j-1)^2(2j-1)(2j+3)}P_{j+1}(\cos\psi) , \qquad (11.42)$$

$$M_6(\cos\psi) := \sum_{j=2}^{\infty} \frac{3(j+1)^2(2j+1)}{(j-1)^2(2j-1)(2j+3)} P_j(\cos\psi), \qquad (11.43)$$

$$M_7(\cos\psi) := \sum_{j=2}^{\infty} \frac{3j(2j+1)}{(j-1)^2(2j-1)} P_j(\cos\psi). \qquad (11.44)$$

11.4 Spatial forms of functions $M_i(\cos\psi)$

We shall now to sum up the infinite sums for $M_i(\cos\psi)$. Simple manipulations with eqns.(11.38)–(11.44) result in fact that $M_i(\cos\psi)$ can be, besides others, expressed in terms of sums

$$\sum_{j=3}^{\infty} \frac{P_j(\cos\psi)}{2j+1} \qquad \text{and/or} \qquad \sum_{j=3}^{\infty} \frac{P_j(\cos\psi)}{(j-1)^2}. \qquad (11.45)$$

The first sum can be expressed in full elliptic integrals (Pick et al., 1973, Appendix 18) which can be calculated only approximately by a method of numerical quadrature (Press et al., 1992, sect.6.11). The second sum is equal to a definite integral, the primitive function of which cannot be expressed in a closed analytical form (Pick et al., 1973, Appendix 18) but again only numerically. We can thus see that sums (11.38)–(11.44) cannot be expressed in closed analytical forms. Therefore, our method of summation will be based on the following idea. Since the kernels $M_i(\cos\psi)$ are singular at the point $\psi = 0$, we shall remove the contributions that are responsible for the singular behaviour at the point $\psi = 0$ from sums (11.38)–(11.44). These contributions will be expressed in closed analytical forms. Having removed singular contributions, the rest will be represented by quickly convergent infinite series, which are bounded on the whole interval $0 \le \psi \le \pi$. Prescribing an error of computation, they can be simply summed up numerically.

Let us start summing up infinite series (11.38). Shifting the summation index to $j - 1$, the fraction occurring in this series can be decomposed as

$$\frac{1}{(j-2)j(2j+1)} = \frac{4j+1}{2(j^2-1)(2j-1)(2j+3)} + \frac{3(6j^3-j^2-2j+2)}{2j(j-2)(j^2-1)(4j^2-1)(2j+3)}. \qquad (11.46)$$

By means of the last equation and eqn.(9.62), function $M_1(\cos\psi)$ reads

$$M_1(\cos\psi) = -\frac{3}{4}\left[\ln\left(\sin\frac{\psi}{2}+\sin^2\frac{\psi}{2}\right)+\frac{6}{7}\cos\psi-\frac{1}{10}\right]+R_1(\cos\psi), \qquad (11.47)$$

where

$$R_1(\cos\psi) = \frac{9}{2}\sum_{j=3}^{\infty} \frac{6j^3-j^2-2j+2}{j(j-2)(j^2-1)(4j^2-1)(2j+3)}\frac{dP_j(\cos\psi)}{d\cos\psi}. \qquad (11.48)$$

Let us prove that the function $R_1(\cos\psi)$ is bounded for $\psi \in \langle 0,\pi\rangle$. For those ψ's, the derivatives of Legendre polynomials can be estimated by eqn.(9.66) which yields

$$|R_1(\cos\psi)| \le \frac{9}{2}\sum_{j=3}^{\infty}\frac{j(6j^3 - j^2 - 2j + 2)}{(j-2)(j^2-1)(4j^2-1)(2j+3)} <$$

$$< \frac{9}{2}\sum_{j=3}^{\infty}\frac{6j^3 - j^2 - 2j + 2}{(j-2)(j-1)(4j^2-1)(2j+3)} < \qquad (11.49)$$

$$< \frac{9}{2}\sum_{j=3}^{\infty}\frac{6j^3 - j^2 - 2j + 2 + (j+19)}{(j-2)(j-1)(4j^2-1)(2j+3)} = \frac{9}{2}\sum_{j=3}^{\infty}\frac{3j^2 - 5j + 7}{(j-2)(j-1)(4j^2-1)} <$$

$$< \frac{9}{8}\sum_{j=3}^{\infty}\frac{12j^2 - 20j + 28 + (20j - 31)}{(j-2)(j-1)(4j^2-1)} = \frac{27}{8}\sum_{j=3}^{\infty}\frac{1}{(j-2)(j-1)} = \frac{27}{8},$$

where we have used eqn.(10.55). It should be pointed out that the last estimate
is rather weak; Figure 11.2 shows that $|R_1(\cos\psi)| < 1.2$.

Nevertheless, it is important that the function $R_1(\cos\psi)$ is bounded at any
point in the interval $0 \le \psi \le \pi$, so that the singularity of function $M_1(\cos\psi)$ at
the point $\psi = 0$ is expressed analytically by the first term on the right-hand side of
eqn.(11.47). We can see that $M_1(\cos\psi)$ has a logarithmic singularity at the point
$\psi = 0$. In addition, function $R_1(\cos\psi)$ can simply be computed numerically since
it is represented by a quickly convergent series. Figure 11.1 demonstrates this fact
in a transparent way; for $\psi = 0$ it plots the decay of magnitudes of series terms
in eqn.(11.48) with increasing degree j. Inspecting Figure 11.1 we can estimate
that it is sufficient to sum up the infinite series (11.48) for $R_1(\cos\psi)$ up to $j \approx 25$
in order to achieve an absolute accuracy of the order of 0.01. This accuracy is
sufficient for computing the 'spherical-ellipsoidal' Stokes function $S^{\text{elco}}(\Omega, \Omega')$ in
the frame of the $O(e_0^2)$-approximation.

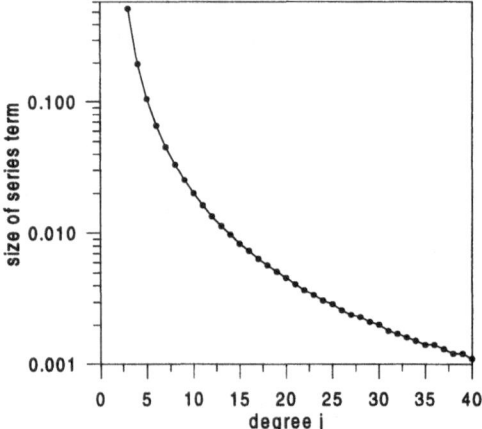

Figure 11.1: The sizes of the particular terms creating the infinite series for
$R_1(1)$.

The spatial forms of the other integral kernel $M_i(\cos\psi)$, $i = 2, ..., 7$, can be expressed in a similar fashion as the kernel $M_1(\cos\psi)$; after some cumbersome, but straightforward algebra we can arrive at

$$M_2(\cos\psi) = \frac{\cos\frac{\psi}{2}}{2\sin\frac{\psi}{2}} + \frac{\cos\frac{\psi}{2}(6 + 5\sin\frac{\psi}{2})}{2(1 + \sin\frac{\psi}{2})} - \frac{33}{8}\sin\psi\ln\left(\sin\frac{\psi}{2} + \sin^2\frac{\psi}{2}\right) -$$
$$- \frac{207}{80}\sin\psi - \frac{181}{28}\sin\psi\cos\psi + R_2(\cos\psi),$$

$$M_3(\cos\psi) = -\frac{3}{4}\left(3\cos^2\psi - 1\right)\ln\left(\sin\frac{\psi}{2} + \sin^2\frac{\psi}{2}\right) - \frac{3}{2}(1 + 3\cos\psi)\sin\frac{\psi}{2} +$$
$$+ \frac{9}{8} + \frac{3}{2}\cos\psi - \frac{21}{8}\cos^2\psi + R_3(\cos\psi),$$

$$M_4(\cos\psi) = -\frac{3}{4}\left[\ln\left(\sin\frac{\psi}{2} + \sin^2\frac{\psi}{2}\right) + \frac{6}{7}\cos\psi - \frac{1}{10}\right] +$$
$$+ R_4(\cos\psi), \tag{11.50}$$

$$M_5(\cos\psi) = R_5(\cos\psi),$$

$$M_6(\cos\psi) = -\frac{3}{2}\left[\cos\psi\ln\left(\sin\frac{\psi}{2} + \sin^2\frac{\psi}{2}\right) + 2\sin\frac{\psi}{2} + \cos\psi - 1\right] +$$
$$+ R_6(\cos\psi),$$

$$M_7(\cos\psi) = -3\left[\cos\psi\ln\left(\sin\frac{\psi}{2} + \sin^2\frac{\psi}{2}\right) + 2\sin\frac{\psi}{2} + \cos\psi - 1\right] +$$
$$+ R_7(\cos\psi),$$

where the residuals $R_i(\cos\psi)$ are of the forms

$$R_2(\cos\psi) = \frac{3}{4}\sin\psi\sum_{j=3}^{\infty}\frac{138j^3 + 17j^2 - 6j + 16}{(j-2)j(j^2-1)(4j^2-1)(2j+3)}\frac{dP_j(\cos\psi)}{d\cos\psi},$$

$$R_3(\cos\psi) = \frac{3}{2}\sum_{j=3}^{\infty}\frac{1}{(j-2)(2j+1)}P_j(\cos\psi),$$

$$R_4(\cos\psi) = \frac{3}{2}\sum_{j=3}^{\infty}\frac{84j^4 - 132j^3 - 15j^2 + 72j + 6}{(j-2)^2(j^2-1)(4j^2-1)(4j^2-9)}\frac{dP_j(\cos\psi)}{d\cos\psi}, \tag{11.51}$$

$$R_5(\cos\psi) = 3\sum_{j=3}^{\infty}\frac{j(2j-1)}{(j-2)^2(2j-3)(2j+1)}P_j(\cos\psi),$$

$$R_6(\cos\psi) = \frac{3}{2}\sum_{j=2}^{\infty}\frac{10j^2 + 15j - 1}{(j-1)^2(2j-1)(2j+3)}P_j(\cos\psi),$$

$$R_7(\cos\psi) = 3\sum_{j=2}^{\infty}\frac{4j-1}{(j-1)^2(2j-1)}P_j(\cos\psi).$$

Figure 11.2 plots the residuals $R_i(\cos\psi)$, $i = 1, ..., 7$, within the interval $0 \leq \psi \leq \pi$. We can observe that $R_i(\cos\psi)$ are 'sufficiently' smooth functions bounded for all angles ψ's. This is a consequence of the fact that the magnitudes of series

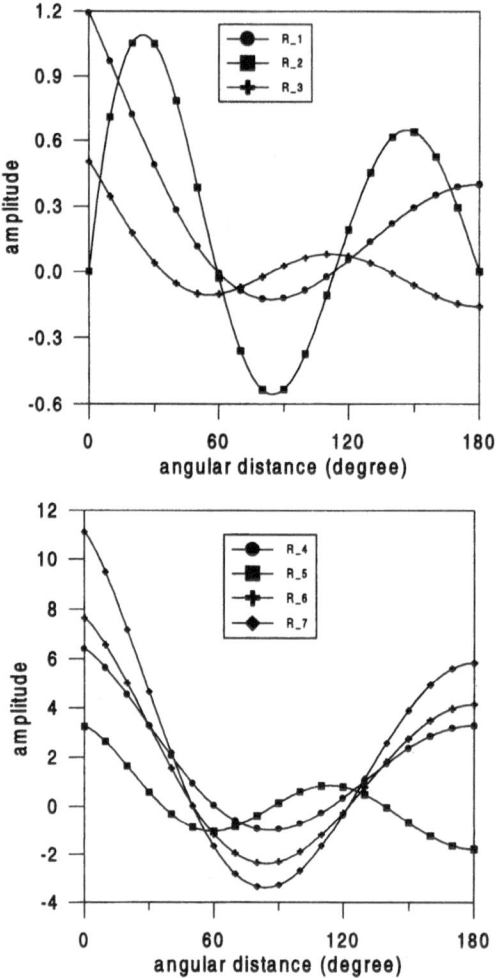

Figure 11.2: Functions $R_i(\cos\psi)$, i=1,...,7, for $\psi \in <0,\pi>$.

terms in infinite sums (11.51) quickly decrease with increasing summation index
j. In order to achieve an absolute accuracy of the order of 0.01, which is sufficient
in the framework of $O(e_0^2)$-approximation, the series (11.51) may be truncated at
degree of about $j \approx 25$.

The above formulae make it possible for us to study the behaviour of functions
$M_i(\cos\psi)$ in the vicinity of the point $\psi = 0$. We can readily see that

$$\lim_{\psi\to 0} M_2(\cos\psi) \quad \sim \quad \frac{1}{\psi}\,, \tag{11.52}$$

$$\lim_{\psi\to 0} M_i(\cos\psi) \quad \sim \quad \ln(\psi/2) \qquad \text{if } i = 1,3,4,6,7\,, \tag{11.53}$$

$$\left|\lim_{\psi\to 0} M_5(\cos\psi)\right| \quad < \quad A \qquad (A \text{ is finite})\,. \tag{11.54}$$

Moreover, taking formula (C.11.18) into account, we can see that

$$2 \sin \vartheta \cos \vartheta \sin \psi \cos \psi \cos \alpha|_{\vartheta = \vartheta'} = \sin^2 \psi \left(\cos^2 \vartheta - \sin^2 \vartheta \cos^2 \alpha \right) . \qquad (11.55)$$

By this, the function $h_2(\vartheta, \psi, \alpha)$ defined by eqn.(11.32) behaves like ψ when $\psi \to 0$. Consequently, 'spherical-ellipsoidal' Stokes function $S^{\text{elco}}(\Omega, \Omega')$ has a logarithmic singularity at the point $\psi = 0$, i.e., $S^{\text{elco}}(\Omega, \Omega') \sim \ln(\psi/2)$ for $\psi \to 0$. Using eqns.(C.11.25) and (C.11.26), we can get the proportional factor of this singularity:

$$S^{\text{elco}}(\Omega, \Omega') \sim 3 \sin^2 \vartheta \ln(\psi/2) \qquad \text{for } \psi \to 0 . \qquad (11.56)$$

This also means that the singularity of $S^{\text{elco}}(\Omega, \Omega')$ at the point $\psi = 0$ is weaker than that of the spherical Stokes function.

11.5   Conclusion

This work was motivated by the question whether the solution to Stokes's boundary-value problem with ellipsoidal corrections involved in boundary condition for anomalous gravity can be expressed in a closed spatial form which would be suitable for numerical computations. The objective behind this effort is to avoid applying the iterative approach usually recommended for solving this type of geodetic boundary-value problem. To answer this question, we first constructed Green's function in terms of spherical harmonics. The spherical harmonic coefficients of sought potential satisfy an infinite system of linear algebraic equations with a tridiagonal system matrix. We found the solution to this system with the accuracy of the order of the Earth's eccentricity. This is an acceptable approximation since it is consistent with the linearization errors hidden behind the formulated boundary-value problem. Moreover, this accuracy is still fairly good for today's requirement on geoid height computations. Within this accuracy, we have shown that the solution can be written as a surface integral taken over the full solid angle with integration kernel consisting of the traditional spherical Stokes function and the correction due to the appearance of the ellipsoidal corrections in the boundary conditions for anomalous gravity. We have managed to express the correcting integration kernel, originally represented in the form of an infinite spherical harmonics, as a finite combination of elementary analytical functions exactly representing the behaviour of integration kernel at the vicinity of its singular point $\psi = 0$. The important conclusion is that the correcting integration kernel has only a logarithmic singularity at $\psi = 0$ that is weaker than the singularity of the spherical Stokes function.

A.11   Spectral form of ellipsoidal corrections

In this section we aim to derive the spectral representation of ellipsoidal correction terms, $\varepsilon_h(T)$ and $\varepsilon_\gamma(T)$, defined by eqns.(1.30) and (1.40). Substituting spherical harmonic expansion (11.5) of the potential T into these equations, taking the derivative of T with respect to ϑ for the term $\varepsilon_h(T)$, and putting $r = R$, we get

$$\varepsilon_h(T) = \frac{e_0^2}{R} \sum_{\substack{j=0 \\ j \neq 1}}^{\infty} \sum_{m=-j}^{j} T_{jm} \sin\vartheta\cos\vartheta \frac{\partial Y_{jm}(\Omega)}{\partial\vartheta} , \qquad (A.11.1)$$

$$\varepsilon_\gamma(T) = \frac{e_0^2}{R} \sum_{\substack{j=0 \\ j \neq 1}}^{\infty} \sum_{m=-j}^{j} T_{jm} \left(3\cos^2\vartheta - 2\right) Y_{jm}(\Omega) . \qquad (A.11.2)$$

To express the product of derivative $\partial Y_{jm}(\Omega)/\partial\vartheta$ with $\sin\vartheta\cos\vartheta$ in terms of $Y_{jm}(\Omega)$, we use the recurrence formula for the first derivative of spherical harmonics (Varshalovich et al., 1989, sect.5.8.2):

$$\sin\vartheta\cos\vartheta\frac{\partial Y_{jm}(\Omega)}{\partial\vartheta} = j\sqrt{\frac{(j+1)^2 - m^2}{(2j+1)(2j+3)}}\cos\vartheta\, Y_{j+1,m}(\Omega)-$$

$$-(j+1)\sqrt{\frac{j^2 - m^2}{(2j-1)(2j+1)}}\cos\vartheta\, Y_{j-1,m}(\Omega) . \qquad (A.11.3)$$

By recurrence formulæ (ibid., p.145, eqn.(2)),

$$\cos\vartheta\, Y_{j+1,m}(\Omega) = \sqrt{\frac{(j+2)^2 - m^2}{(2j+3)(2j+5)}} Y_{j+2,m}(\Omega) + \sqrt{\frac{(j+1)^2 - m^2}{(2j+1)(2j+3)}} Y_{jm}(\Omega) , \qquad (A.11.4)$$

and

$$\cos\vartheta\, Y_{j-1,m}(\Omega) = \sqrt{\frac{(j-1)^2 - m^2}{(2j-3)(2j-1)}} Y_{j-2,m}(\Omega) + \sqrt{\frac{j^2 - m^2}{(2j-1)(2j+1)}} Y_{jm}(\Omega) , \qquad (A.11.5)$$

we get

$$\sin\vartheta\cos\vartheta\frac{\partial Y_{jm}(\Omega)}{\partial\vartheta} = -\frac{j+1}{2j-1}\sqrt{\frac{[(j-1)^2 - m^2][j^2 - m^2]}{(2j-3)(2j+1)}} Y_{j-2,m}(\Omega)+$$

$$+\frac{3m^2 - j(j+1)}{(2j-1)(2j+3)} Y_{jm}(\Omega) + \frac{j}{2j+3}\sqrt{\frac{[(j+1)^2 - m^2][(j+2)^2 - m^2]}{(2j+1)(2j+5)}} Y_{j+2,m}(\Omega) . \qquad (A.11.6)$$

To obtain a similar expression for the term $\varepsilon_\gamma(T)$, we make use of the recurrence formula for spherical harmonics (ibid., p.145, eqn.(5)) and get

$$\left(3\cos^2\vartheta - 2\right) Y_{jm}(\Omega) = \frac{3}{2j-1}\sqrt{\frac{[(j-1)^2 - m^2][j^2 - m^2]}{(2j-3)(2j+1)}} Y_{j-2,m}(\Omega)-$$

$$- \left[\frac{2[3m^2 - j(j+1)]}{(2j-1)(2j+3)} + 1 \right] Y_{jm}(\Omega) + \tag{A.11.7}$$

$$+ \frac{3}{2j+3} \sqrt{\frac{[(j+1)^2 - m^2][(j+2)^2 - m^2]}{(2j+1)(2j+5)}} Y_{j+2,m}(\Omega) .$$

Finally, substitution of eqns.(A.11.6) and (A.11.7) into (A.11.1) and (A.11.2) results in spherical harmonic representation of the sum of ellipsoidal correction terms,

$$
\begin{aligned}
\varepsilon_h(T) + \varepsilon_\gamma(T) = \frac{e_0^2}{R} \sum_{\substack{j=0 \\ j \neq 1}}^{\infty} \sum_{m=-j}^{j} & \left\{ \frac{j+1}{2j-1} \sqrt{\frac{[(j-1)^2 - m^2][j^2 - m^2]}{(2j-3)(2j+1)}} T_{j-2,m} + \right. \\
& + \left[\frac{j(j+1) - 3m^2}{(2j-1)(2j+3)} - 1 \right] T_{jm} - \\
& \left. - \frac{j}{2j+3} \sqrt{\frac{[(j+1)^2 - m^2][(j+2)^2 - m^2]}{(2j+1)(2j+5)}} T_{j+2,m} \right\} Y_{jm}(\Omega) .
\end{aligned}
\tag{A.11.8}
$$

B.11 An approximate solution to tridiagonal system of equations

The set of equations to be solved is

$$
\begin{pmatrix}
1+b_1 & c_1 & 0 & \cdots & & & \\
a_2 & 1+b_2 & c_2 & \cdots & & & \\
& & \cdots & & & & \\
& & \cdots & a_{n-1} & 1+b_{n-1} & c_{n-1} \\
& & \cdots & 0 & a_n & 1+b_n
\end{pmatrix}
\times
$$

$$
\times
\begin{pmatrix}
u_1 \\
u_2 \\
\cdots \\
u_{n-1} \\
u_n
\end{pmatrix}
=
\begin{pmatrix}
r_1 \\
r_2 \\
\cdots \\
r_{n-1} \\
r_n
\end{pmatrix} .
\tag{B.11.1}
$$

The solution to this system of equations with tridiagonal matrix can be carried out by the following recurrence formulae (Isaacson and Keller, 1973, sect.2.3.2). For $j = 2, ..., n$, let us generate the quantities:

$$\gamma_j := \frac{c_{j-1}}{\beta_{j-1}} , \tag{B.11.2}$$

$$\beta_j := 1 + b_j - a_j \gamma_j , \tag{B.11.3}$$

$$\alpha_j := \frac{r_j - a_j \alpha_{j-1}}{\beta_j} , \tag{B.11.4}$$

with starting values

$$\beta_1 := 1 + b_1 , \qquad\qquad \alpha_1 := \frac{r_1}{\beta_1} . \qquad (B.11.5)$$

Having created α_j, β_j, and γ_j, the solution to tridiagonal system (B.11.1) is

$$u_n = \alpha_n \qquad\qquad (B.11.6)$$
$$u_j = \alpha_j - \gamma_{j+1} u_{j+1} \qquad \text{for } j = n-1, ..., 1 . \qquad (B.11.7)$$

Now, let us suppose that the elements a_j, b_j, and c_j are small quantities such that $|a_j| \le \varepsilon$, $|b_j| \le \varepsilon$, $|c_j| \le \varepsilon$, and $\varepsilon \ll 1$. To get an approximate solution of eqns.(B.11.1) with the accuracy of the order of ε, we can approximately put

$$\beta_j \doteq 1 + b_j + O(\varepsilon^2) , \qquad (B.11.8)$$

since $|\gamma_j| \le \varepsilon$. Within the same accuracy, eqn.(B.11.4) can be replaced by

$$\alpha_j \doteq r_j - b_j r_j - a_j r_{j-1} + O(\varepsilon^2) . \qquad (B.11.9)$$

Substituting these formulae to eqns.(B.11.6) and (B.11.7) leads to the approximate solution to eqn.(B11.1)

$$\boxed{u_j \doteq r_j - b_j r_j - a_j r_{j-1} - c_j r_{j+1} + O(\varepsilon^2) , \qquad j = 1, 2, ..., n ,} \qquad (B.11.10)$$

assuming that $c_n = 0$.

C.11   Different forms of the addition theorem for spherical harmonics

In this section, we shall derive different forms of addition theorem for spherical harmonics. Let us start with the recurrence relation for the spherical harmonics (Varshalovich et al., 1989, sect.5.8.2)

$$\sqrt{\frac{2j+3}{2j+1}} [(j+1)^2 - m^2] Y_{jm}(\Omega) = \sin^2 \vartheta \, \frac{\partial Y_{j+1,m}(\Omega)}{\partial \cos \vartheta} + (j+1) \cos \vartheta \, Y_{j+1,m}(\Omega) .$$
$$(C.11.1)$$

This recurrence formula can be rewritten in a different form:

$$\sqrt{\frac{2j+3}{2j+5}} [(j+2)^2 - m^2] Y_{j+2,m}(\Omega) = -\sin^2 \vartheta \, \frac{\partial Y_{j+1,m}(\Omega)}{\partial \cos \vartheta} + (j+2) \cos \vartheta \, Y_{j+1,m}(\Omega) .$$
$$(C.11.2)$$

Multiplying eqn.(C.11.1) by complex conjugate relation to (C.11.2) taken at a point Ω', we get

$$(2j+3)\sqrt{\frac{[(j+1)^2 - m^2][(j+2)^2 - m^2]}{(2j+1)(2j+5)}} Y_{jm}(\Omega) Y^*_{j+2,m}(\Omega') =$$

$$
\begin{aligned}
= {} & -\sin^2 \vartheta \sin^2 \vartheta' \, \frac{\partial Y_{j+1,m}(\Omega)}{\partial \cos \vartheta} \frac{\partial Y^*_{j+1,m}(\Omega')}{\partial \cos \vartheta'} - \\
& - (j+1) \cos \vartheta \sin^2 \vartheta' \, Y_{j+1,m}(\Omega) \frac{\partial Y^*_{j+1,m}(\Omega')}{\partial \cos \vartheta'} + \\
& + (j+2) \sin^2 \vartheta \cos \vartheta' \, Y^*_{j+1,m}(\Omega') \frac{\partial Y_{j+1,m}(\Omega)}{\partial \cos \vartheta} + \\
& + (j+1)(j+2) \cos \vartheta \cos \vartheta' \, Y_{j+1,m}(\Omega) Y^*_{j+1,m}(\Omega') \, .
\end{aligned}
\tag{C.11.3}
$$

Summing eqn.(C.11.3) from $m = -j - 1$ up to $m = j + 1$, and realizing that the factor $(j+1)^2 - m^2$ is equal to zero for $m = \pm(j+1)$, we get

$$
(2j+3) \sum_{m=-j}^{j} \sqrt{\frac{[(j+1)^2 - m^2][(j+2)^2 - m^2]}{(2j+1)(2j+5)}} \, Y_{jm}(\Omega) Y^*_{j+2,m}(\Omega') =
$$

$$
\begin{aligned}
= {} & -\sin^2 \vartheta \sin^2 \vartheta' \sum_{m=-(j+1)}^{j+1} \frac{\partial Y_{j+1,m}(\Omega)}{\partial \cos \vartheta} \frac{\partial Y^*_{j+1,m}(\Omega')}{\partial \cos \vartheta'} - \\
& - (j+1) \cos \vartheta \sin^2 \vartheta' \sum_{m=-(j+1)}^{j+1} Y_{j+1,m}(\Omega) \frac{\partial Y^*_{j+1,m}(\Omega')}{\partial \cos \vartheta'} + \\
& + (j+2) \sin^2 \vartheta \cos \vartheta' \sum_{m=-(j+1)}^{j+1} Y^*_{j+1,m}(\Omega') \frac{\partial Y_{j+1,m}(\Omega)}{\partial \cos \vartheta} + \\
& + (j+1)(j+2) \cos \vartheta \cos \vartheta' \sum_{m=-(j+1)}^{j+1} Y_{j+1,m}(\Omega) Y^*_{j+1,m}(\Omega') \, .
\end{aligned}
\tag{C.11.4}
$$

The sums of the products of spherical harmonics and their derivatives will be simplified by means of the Laplace addition theorem for spherical harmonics. Taking the Laplace addition theorem (11.20) for index $j + 1$, we have

$$
\frac{2j+3}{4\pi} P_{j+1}(\cos \psi) = \sum_{m=-(j+1)}^{j+1} Y_{j+1,m}(\Omega) Y^*_{j+1,m}(\Omega') \, .
\tag{C.11.5}
$$

Differentiating eqn.(C.11.5) with respect to $\cos \vartheta$ and $\cos \vartheta'$, and substituting the results to eqn.(C.11.4), we get

$$
4\pi \sum_{m=-j}^{j} \sqrt{\frac{[(j+1)^2 - m^2][(j+2)^2 - m^2]}{(2j+1)(2j+5)}} \, Y_{jm}(\Omega) Y^*_{j+2,m}(\Omega') =
$$

$$
\begin{aligned}
= {} & -\sin^2 \vartheta \sin^2 \vartheta' \, \frac{\partial^2 P_{j+1}(\cos \psi)}{\partial \cos \vartheta \, \partial \cos \vartheta'} - (j+1) \cos \vartheta \sin^2 \vartheta' \, \frac{\partial P_{j+1}(\cos \psi)}{\partial \cos \vartheta'} + \\
& + (j+2) \sin^2 \vartheta \cos \vartheta' \, \frac{\partial P_{j+1}(\cos \psi)}{\partial \cos \vartheta} + (j+1)(j+2) \cos \vartheta \cos \vartheta' \, P_{j+1}(\cos \psi) \, .
\end{aligned}
\tag{C.11.6}
$$

The partial derivatives of Legendre polynomials $P_{j+1}(\cos\psi)$ with respect to $\cos\vartheta$ and $\cos\vartheta'$ will now be expressed in terms of the ordinary derivatives of the Legendre polynomials with respect to $\cos\psi$ by making use well-known formulae of spherical trigonometry (e.g., Heiskanen and Moritz, 1967, eqn.(2-208)),

$$\cos\psi \;=\; \cos\vartheta\cos\vartheta' + \sin\vartheta\sin\vartheta'\cos(\lambda-\lambda')\,, \qquad (C.11.7)$$

$$\sin\psi\cos\alpha \;=\; \sin\vartheta\cos\vartheta' - \cos\vartheta\sin\vartheta'\cos(\lambda-\lambda')\,, \qquad (C.11.8)$$

$$\sin\psi\sin\alpha \;=\; -\sin\vartheta'\sin(\lambda-\lambda')\,, \qquad (C.11.9)$$

where α is the azimuth between directions Ω and Ω'. Realizing that

$$\frac{\partial P_{j+1}(\cos\psi)}{\partial\cos\vartheta} = \frac{dP_{j+1}(\cos\psi)}{d\cos\psi}\frac{\partial\cos\psi}{\partial\cos\vartheta}\,, \qquad (C.11.10)$$

and taking the derivative of eqn.(C.11.7) with respect to $\cos\vartheta$, we have

$$\frac{\partial P_{j+1}(\cos\psi)}{\partial\cos\vartheta} = [\cos\vartheta' - \cot\vartheta\sin\vartheta'\cos(\lambda-\lambda')]\frac{dP_{j+1}(\cos\psi)}{d\cos\psi}\,. \qquad (C.11.11)$$

Similarly, we can write

$$\frac{\partial P_{j+1}(\cos\psi)}{\partial\cos\vartheta'} = [\cos\vartheta - \cot\vartheta'\sin\vartheta\cos(\lambda-\lambda')]\frac{dP_{j+1}(\cos\psi)}{d\cos\psi}\,. \qquad (C.11.12)$$

Furthermore, differentiating eqn.(C.11.11) with respect to $\cos\vartheta'$, we obtain the 2nd-order derivative occurring in eqn.(C.11.6),

$$\frac{\partial^2 P_{j+1}(\cos\psi)}{\partial\cos\vartheta\,\partial\cos\vartheta'} = [1 + \cot\vartheta\cot\vartheta'\cos(\lambda-\lambda')]\frac{dP_{j+1}(\cos\psi)}{d\cos\psi} +$$

$$+\, [\cos\vartheta' - \cot\vartheta\sin\vartheta'\cos(\lambda-\lambda')]\,[\cos\vartheta - \cot\vartheta'\sin\vartheta\cos(\lambda-\lambda')]\frac{d^2 P_{j+1}(\cos\psi)}{d(\cos\psi)^2}\,. $$
$$(C.11.13)$$

Substituting eqns.(C.11.11)–(C.11.13) into eqn.(C.11.6), we have

$$4\pi\sum_{m=-j}^{j}\sqrt{\frac{[(j+1)^2 - m^2][(j+2)^2 - m^2]}{(2j+1)(2j+5)}}\,Y_{jm}(\Omega)Y^*_{j+2,m}(\Omega') =$$

$$= \; -\sin\vartheta\sin\vartheta'\,[\sin\vartheta\cos\vartheta' - \cos\vartheta\sin\vartheta'\cos(\lambda-\lambda')]\,\times$$

$$\times\,[\sin\vartheta'\cos\vartheta - \cos\vartheta'\sin\vartheta\cos(\lambda-\lambda')]\frac{d^2 P_{j+1}(\cos\psi)}{d(\cos\psi)^2} - \qquad (C.11.14)$$

$$-\;\;\sin\vartheta\sin\vartheta'\,[\sin\vartheta\sin\vartheta' + \cos\vartheta\cos\vartheta'\cos(\lambda-\lambda')]\frac{dP_{j+1}(\cos\psi)}{d\cos\psi} -$$

$$-\;\;(j+1)\cos\vartheta\sin\vartheta'\,[\sin\vartheta'\cos\vartheta - \cos\vartheta'\sin\vartheta\cos(\lambda-\lambda')]\frac{dP_{j+1}(\cos\psi)}{d\cos\psi} +$$

$$+\;\;(j+2)\sin\vartheta\cos\vartheta'\,[\sin\vartheta\cos\vartheta' - \cos\vartheta\sin\vartheta'\cos(\lambda-\lambda')]\frac{dP_{j+1}(\cos\psi)}{d\cos\psi} +$$

$$+\;\;(j+1)(j+2)\cos\vartheta\cos\vartheta'\,P_{j+1}(\cos\psi)\,. $$

The functions in square brackets standing in front of the derivatives of Legendre polynomials will be expressed by means of the angular distance ψ and the azimuth α between directions Ω and Ω'. Using eqns.(C.11.7)–(C.11.9), we can, after some algebraic manipulations, obtain:

$$\sin\vartheta'\cos\vartheta - \cos\vartheta'\sin\vartheta\cos(\lambda-\lambda') = \frac{1}{\sin\vartheta'}\left(\cos\vartheta - \cos\vartheta'\cos\psi\right) , \quad (C.11.15)$$

$$\sin\vartheta\sin\vartheta' + \cos\vartheta\cos\vartheta'\cos(\lambda-\lambda') = \frac{1}{\sin\vartheta'}\left(\sin\vartheta - \cos\vartheta'\sin\psi\cos\alpha\right) . \quad (C.11.16)$$

Considering eqns.(C.11.8), (C.11.15) and (C.11.16) in eqn.(C.11.14), we have

$$
4\pi \sum_{m=-j}^{j} \sqrt{\frac{[(j+1)^2 - m^2][(j+2)^2 - m^2]}{(2j+1)(2j+5)}}\, Y_{jm}(\Omega)Y_{j+2,m}^{*}(\Omega') =
$$

$$
\begin{aligned}
= & -\sin\vartheta\sin\psi\cos\alpha\,(\cos\vartheta - \cos\vartheta'\cos\psi)\frac{d^2 P_{j+1}(\cos\psi)}{d(\cos\psi)^2} - \\
& -\sin\vartheta\,(\sin\vartheta - \cos\vartheta'\sin\psi\cos\alpha)\frac{dP_{j+1}(\cos\psi)}{d\cos\psi} - \\
& -(j+1)\cos\vartheta\,(\cos\vartheta - \cos\vartheta'\cos\psi)\frac{dP_{j+1}(\cos\psi)}{d\cos\psi} + \\
& +(j+2)\sin\vartheta\cos\vartheta'\sin\psi\cos\alpha\,\frac{dP_{j+1}(\cos\psi)}{d\cos\psi} + \\
& +(j+1)(j+2)\cos\vartheta\cos\vartheta'\,P_{j+1}(\cos\psi) .
\end{aligned}
\tag{C.11.17}
$$

Multiplying eqn.(C.11.7) by $\cos\vartheta/\sin\vartheta$ and adding the result to eqn.(C.11.8), we get

$$\cos\vartheta' = \cos\vartheta\cos\psi + \sin\vartheta\sin\psi\cos\alpha . \tag{C.11.18}$$

Employing the last formula and Legendre's differential equation,

$$\sin^2\psi\,\frac{d^2 P_{j+1}(\cos\psi)}{d(\cos\psi)^2} - 2\cos\psi\,\frac{dP_{j+1}(\cos\psi)}{d\cos\psi} + (j+1)(j+2)P_{j+1}(\cos\psi) = 0 , \tag{C.11.19}$$

the sum of the first, the second and the last term on the right-hand side of eqn.(C.11.17) reads

$$
\begin{aligned}
& -\sin\vartheta\sin\psi\cos\alpha\,(\cos\vartheta - \cos\vartheta'\cos\psi)\frac{d^2 P_{j+1}(\cos\psi)}{d(\cos\psi)^2} - \\
& -\sin\vartheta\,(\sin\vartheta - \cos\vartheta'\sin\psi\cos\alpha)\frac{dP_{j+1}(\cos\psi)}{d\cos\psi} + \\
& +(j+1)(j+2)\cos\vartheta\cos\vartheta'\,P_{j+1}(\cos\psi) = \\
= & -\sin\vartheta\,(\sin\vartheta + \cos\vartheta\sin\psi\cos\psi\cos\alpha - \\
& -\sin\vartheta\cos^2\psi\cos^2\alpha - \sin\vartheta\cos^2\alpha)\frac{dP_{j+1}(\cos\psi)}{d\cos\psi} + \\
& +(j+1)(j+2)\left(\cos^2\vartheta\cos\psi + 2\sin\vartheta\cos\vartheta\sin\psi\cos\alpha - \right. \\
& \left. -\sin^2\vartheta\cos\psi\cos^2\alpha\right)P_{j+1}(\cos\psi) .
\end{aligned}
\tag{C.11.20}
$$

Moreover, by formula (C.11.18), the sum od the third and fourth term on the right-hand side of eqn.(C.11.17) becomes

$$
- (j+1)\cos\vartheta\,(\cos\vartheta - \cos\vartheta'\cos\psi)\frac{dP_{j+1}(\cos\psi)}{d\cos\psi} +
$$

$$
+ (j+2)\sin\vartheta\cos\vartheta'\sin\psi\cos\alpha\,\frac{dP_{j+1}(\cos\psi)}{d\cos\psi} =
$$

$$
= -(j+1)\sin\psi\left(\cos^2\vartheta\sin\psi - 2\sin\vartheta\cos\vartheta\cos\psi\cos\alpha - \right. \tag{C.11.21}
$$

$$
- \sin^2\vartheta\sin\psi\cos^2\alpha\Big)\frac{dP_{j+1}(\cos\psi)}{d\cos\psi} +
$$

$$
+ \sin\vartheta\left(\cos\vartheta\sin\psi\cos\psi\cos\alpha + \sin\vartheta\sin^2\psi\cos^2\alpha\right)\frac{dP_{j+1}(\cos\psi)}{d\cos\psi}\ .
$$

Substituting eqns.(C.11.20) and (C.11.21) into eqns.(C.11.17), we finally have

$$
4\pi\sum_{m=-j}^{j}\sqrt{\frac{[(j+1)^2 - m^2]\,[(j+2)^2 - m^2]}{(2j+1)(2j+5)}}\,Y_{jm}(\Omega)Y^*_{j+2,m}(\Omega') =
$$

$$
= \sin^2\vartheta\left(\cos^2\alpha - \sin^2\alpha\right)\frac{dP_{j+1}(\cos\psi)}{d\cos\psi} - \tag{C.11.22}
$$

$$
- (j+1)\sin\psi\left(\cos^2\vartheta\sin\psi - 2\sin\vartheta\cos\vartheta\cos\psi\cos\alpha - \right.
$$

$$
- \sin^2\vartheta\sin\psi\cos^2\alpha\Big)\frac{dP_{j+1}(\cos\psi)}{d\cos\psi} +
$$

$$
+ (j+1)(j+2)\left(\cos^2\vartheta\cos\psi + 2\sin\vartheta\cos\vartheta\sin\psi\cos\alpha - \right.
$$

$$
- \sin^2\vartheta\cos\psi\cos^2\alpha\Big)P_{j+1}(\cos\psi)\ .
$$

In an analogous way, other forms of the addition theorem for spherical harmonics can be derived. We introduce one of them without demonstrating a detailed proof:

$$
4\pi\sum_{m=-j}^{j}\sqrt{\frac{[(j+1)^2 - m^2]\,[(j+2)^2 - m^2]}{(2j+1)(2j+5)}}\,Y_{jm}(\Omega')Y^*_{j+2,m}(\Omega) =
$$

$$
= \sin^2\vartheta\left(\cos^2\alpha - \sin^2\alpha\right)\frac{dP_{j+1}(\cos\psi)}{d\cos\psi} + \tag{C.11.23}
$$

$$
+ (j+2)\sin\psi\left(\cos^2\vartheta\sin\psi - 2\sin\vartheta\cos\vartheta\cos\psi\cos\alpha - \right.
$$

$$
- \sin^2\vartheta\sin\psi\cos^2\alpha\Big)\frac{dP_{j+1}(\cos\psi)}{d\cos\psi} +
$$

$$
+ (j+1)(j+2)\left(\cos^2\vartheta\cos\psi + 2\sin\vartheta\cos\vartheta\sin\psi\cos\alpha - \right.
$$

$$
- \sin^2\vartheta\cos\psi\cos^2\alpha\Big)P_{j+1}(\cos\psi)\ .
$$

Furthermore, the addition theorem for spherical harmonics (B.9.21) now reads

$$\frac{4\pi}{2j+1} \sum_{m=-j}^{j} \left[(j+1)^2 - m^2\right] Y_{jm}(\Omega)Y_{jm}^*(\Omega') = \sin\vartheta \left(\cos\vartheta \sin\psi \cos\psi \cos\alpha - \right.$$

$$- \sin\vartheta \cos^2\psi \cos^2\alpha + \sin\vartheta \sin^2\alpha\right) \frac{dP_{j+1}(\cos\psi)}{d\cos\psi} - \qquad (C.11.24)$$

$$- (j+1)\sin\vartheta \cos\alpha \left(\cos\vartheta \sin\psi - \sin\vartheta \cos\psi \cos\alpha\right) P_{j+1}(\cos\psi) +$$

$$+ (j+1)^2 \left(1 - \sin^2\vartheta \sin^2\alpha\right) P_j(\cos\psi) .$$

Particularly, when $\Omega = \Omega'$, we get

$$4\pi \sum_{m=-j}^{j} \sqrt{\frac{[(j+1)^2 - m^2][(j+2)^2 - m^2]}{(2j+1)(2j+5)}} Y_{jm}(\Omega)Y_{j+2,m}^*(\Omega) =$$

$$= \frac{1}{2}(j+1)(j+2)\left(3\cos^2\vartheta - 1\right) , \qquad (C.11.25)$$

and

$$\frac{4\pi}{2j+1} \sum_{m=-j}^{j} \left[(j+1)^2 - m^2\right] |Y_{jm}(\Omega)|^2 = (j+1)^2 - \frac{1}{2}j(j+1)\sin^2\vartheta . \quad (C.11.26)$$

Chapter 12

The least-squares solution to the discrete altimetry-gravimetry boundary-value problem for determination of the global gravity model

A global gravity model is usually understood to be a representation of the Earth's external gravitational potential in terms of a spherical harmonic series. A theoretically infinite series has to be truncated at some maximum degree since the gravity data used to estimate the series coefficients are not measured continuously in space but are known only at discrete points. The higher the maximum degree, the greater, in principle, is the resolution of the model. By observing the perturbation of a satellite's orbit, low-degree potential coefficients can be estimated. The maximum degree to which the series can be determined from satellite observations depends on the height and inclination distribution of the satellites, as well as the accuracy of the observational instruments. During the last decade, a number of global gravity models have been constructed from orbital analysis (e.g., Reigberg, 1989). The latest satellite global model is given up to degree 50 (Lerch et al., 1992).

To improve the resolution of the satellite models, it is advantageous to combine satellite tracking data with other measurements related to the Earth's gravity field. One obvious quantity, the modulus of the gravity acceleration, results from the absolute and relative gravity measurements performed on the Earth's surface. Combining these measurements with geodetic levelling provides us with information about the gravity potential. When we add the horizontal position of points on the Earth's surface obtained from geodetic positioning, we are equipped with data for computing the global gravity model by solving the boundary-value problem for the Laplace equation in the external space. An obvious advantage of these surface gravity data is their high resolution, which enables us to construct gravity models up to high degrees and orders. Unfortunately, the distribution and the accuracy of the gravity data are not homogeneous over the world. There are still land areas where the gravity data have not been gathered or where the gravity data are not available for political reasons. Furthermore, sea and ocean areas are only partly covered by gravity measurements.

This situation has changed dramatically since satellite altimeter measurements began. The altimeter measures the distance between a satellite and the instantaneous ocean surface. Provided that the position of a satellite is perfectly known, i.e., when the errors in the satellite's orbit have been eliminated (Zlotnicki, 1993; Sailor, 1993), the altimeter data yield direct information about the position of ground points on the ocean's surface. In fact, the ocean's surface deviates from the geoid by actions of the external forces on the layers of the ocean's surface, such as tidal forces, wind stresses or dynamical forces associated with ocean currents. The difference between the sea's surface and the geoid, called the dynamic topography or the sea surface topography (e.g., Wagner, 1989), is mostly represented in terms of low-degree and low-order spherical harmonics. When the dynamic topography is subtracted from the altimeter data, the geoid over the oceans can be considered to be known to an accuracy of about 10 cm.

Additional information pertaining to the gravity and the gravity potential over the continents leads us to ask how to determine the external gravitational field and the figure of the continental surface. This problem, discussed later in this chapter, is the altimetry-gravimetry problem of the first kind (throughout this chapter we will drop the specification 'of the first kind'). In the literature, other types of altimetry-gravimetry problems have been formulated, depending on the boundary conditions prescribed on the oceans (Holota, 1983a,b). Their linearized forms have been treated by several theoreticians (Holota, 1983a,b; Svensson, 1983, 1988; Sansò, 1993). Svensson (1988) proved that a linearized form of the altimetry-gravimetry problem has a non-unique solution. Sacerdote and Sansò (1987) modified the problem by imposing a supplementary constraint on the data and by excluding from the solution the zero-degree harmonic of the anomalous gravitational potential. They proved (later also mentioned by Sansò (1993)) that under such a regularization, the altimetry-gravimetry problem has a unique solution. To ensure the stability of the solution, the anomalous gravitational potential and the altimeter data are to be from the Sobolev space $H^{1/2}$ and the gravity data from the Sobolev space $H^{-1/2}$.

Considering these uniqueness and stability conditions, Svensson (1988) concluded that the classical least-squares solution of the altimetry-gravimetry boundary-value problem is not stable. This statement holds when applied to a continuous case but it is somehow weakened for a discrete least-squares approximation. In the latter case, to estimate a finite spherical harmonic series of the external gravity potential, we may try to use a discrete least-squares method. Therefore, the question arises whether the regularization of the problem by truncating the potential series at a finite cut-off degree is efficient enough to stabilize the discrete least-squares estimation; or, in other words, up to which cut-off degree of the potential series is the discrete least-squares estimation still numerically stable? This chapter attempts to answer this question.

One may ask another question, namely, whether it is useful to include the gravity measurements performed by ships on a part of the sea's surface in the altimetry-gravimetry problem. By requiring the solution of the altimetry-gravimetry problem to take into account these extra data, the problem becomes overdetermined. We will show that the least-squares solution to the overdetermined altimetry-gravimetry problem corresponds to damped least-squares, which are more stable than the original altimetry-gravimetry problem. It should be pointed out that the idea of stabilization of the mixed problem by overlapping the boundary functionals originates from Bjerhammar (1987) but an adequate theory and numerical results have yet to be published.

Up to now, several high-degree global gravity models have been constructed (Rapp, 1986, 1993; Rapp et al., 1991). Their authors use inverse Stokes's formula (Molodenskij et al., 1960) in order to transform the geoidal heights coming from altimeter observations to the gravity anomalies. Numerically, this step is carried out by the least-squares collocation method (Bašić and Rapp, 1991). Unfortunately, the inverse Stokes' formula is represented by a strongly singular surface integral which must be regularized (Rummel, 1977). The fact that this transformation is ill-posed results in the amplification of a high-frequency noise contaminating an altimeter signal. Consequently, some kind of regularization must be taken into account.

The ill-conditioned step of Rapp's method, mainly influencing a high frequency part of the gravity signal, makes the reliability of the high-degree potential coefficients questionable. It is hard to quantify a distortion of these harmonics, although some attempts have been made (Rapp et al., 1991). The aim of this chapter (see also Martinec, 1995) is to present a stabilized method for computing high-degree potential coefficients of a global gravity model without going through the solutions to the ill-conditioned problems. As already introduced, the stabilization will be achieved by cutting high-degree components of the gravitational potential and by overlapping the altimeter data with the gravity observations. The conditionality of the matrix of normal equations will be analysed via the singular value decomposition of the design matrix. Numerical tests will illustrate how both the size of the overlapping area and the truncation of the potential coefficients influence the stability of the least-squares solution.

12.1 Formulation of the boundary-value problem

Let us divide the Earth's surface S into the surface of continents, S_c, and the surface of seas and oceans, S_o, i.e.,

$$S = S_c \cup S_o , \qquad S_c \cap S_o = \emptyset . \qquad (12.1)$$

The altimetry-gravimetry problem (of the first kind) may be formulated as follows (Sansò, 1981; Holota, 1983a,b). Let the horizontal positions of all points on the continental surface S_c be given. Let the gravity potential W_c and the modulus of gravity vector g be surveyed at all the points on S_c. Let the shape of the oceanic surface S_o be known and be identified with the geoid ($W = W_0 = const.$). Determine the radial position of all points on the surface S_c and the gravity potential outside the surface S.

Since it is our intention to deal with the gravity field generated by the Earth's inner masses, we make a few simplifying assumptions. First, we assume that the observations of W_c and g are corrected for the attraction of the atmosphere and the direct gravitational effect of the other bodies, mainly the moon and the sun. Second, we assume that the Earth is a rigid, undeformable body, uniformly rotating (with a constant angular velocity ω) around a fixed axis passing through its center of mass. This assumption excludes consideration of the indirect gravitational effect of other celestial bodies, such as tidal effects. Third, we assume that the actual sea surface is corrected for the dynamical topography induced by the effects of dynamical forces driven by the oceanic currents, wind stresses, or other 'external' forces.

Mathematically, the altimetry-gravimetry problem formulated above is governed by Poisson's equation for the gravity potential W in the space outside S constrained by Dirichlet's boundary conditions on the known boundary S_o, as well as on a radially-unknown boundary S_c, and by the derivative condition on S_c:

$$\nabla^2 W = 2\omega^2 \qquad \text{outside } S\,, \qquad (12.2)$$
$$W = W_0 \qquad \text{on } S_o\,, \qquad (12.3)$$
$$W = W_o \qquad \text{on } S_o\,, \qquad (12.4)$$
$$|\operatorname{grad} W| = g \qquad \text{on } S_c\,, \qquad (12.5)$$
$$W = \tfrac{1}{2}\omega^2 r^2 \sin^2 \vartheta + O\left(\tfrac{1}{r}\right) \qquad r \to \infty\,, \qquad (12.6)$$

where W_0 is the value of potential W on the geoid, and W_c and g are known (measured) functions on S_c. The condition (12.6) excludes the solution to Poisson's equation (12.2) which diverges at infinity. We have introduced the geocentric radius r, the geocentric co-latitude ϑ and longitude λ. The symbol Ω will stand for a pair of angular coordinates ϑ and λ.

To take into consideration the satellite gravity models, we split the gravity potential W into the reference (known) gravity potential U and an anomalous (unknown) gravitational potential T:

$$W = U + T\,. \qquad (12.7)$$

The set of potential coefficients U_{jm} derived from observing perturbations of satellite orbits are taken as the reference. The reference gravity potential U is then represented in the form

$$U(r,\Omega) = \frac{GM}{r}\left[1 + \sum_{j=2}^{j_{ref}}\left(\frac{a}{r}\right)^j \sum_{m=-j}^{j} U_{jm}Y_{jm}(\Omega)\right] + \frac{1}{2}\omega^2 r^2 \sin^2\vartheta , \qquad (12.8)$$

where GM is the geocentric gravitational constant, a is the scaling factor (usually the equatorial radius of the reference ellipsoid), $Y_{jm}(\Omega)$ are complex fully normalized spherical harmonics of degree j and order m, and j_{ref} (≥ 2) is the cut-off degree of the reference potential coefficients.

The non-linear, mixed boundary-value problem (12.2)–(12.6) with a partly fixed and partly free boundary will be reformulated for the anomalous gravitational potential T. We shall treat it in a linearized form in order to make as transparent as possible the difficulty with the correctness of the altimetry-gravimetry problem. Spherical approximation of the linearized altimetry-gravimetry problem is governed by the equations (Sacerdote and Sansò, 1986; Sansò, 1993):

$$\begin{array}{rcll}
\nabla^2 T &=& 0 & \text{outside } S , \qquad (12.9)\\[2mm]
T &=& \delta w & \text{on } S_o , \qquad (12.10)\\[2mm]
-\dfrac{\partial T}{\partial r} - \dfrac{2}{r}T &=& \Delta g & \text{on } S_c , \qquad (12.11)\\[2mm]
T &\sim& O\left(\dfrac{1}{r^{j_{ref}+2}}\right) & r \to \infty , \qquad (12.12)
\end{array}$$

where Δg is the difference between the measured gravity g on the physical surface S_c and the reference gravity $\gamma := |\text{grad}U|$ evaluated on the telluroid (Heiskanen and Moritz, 1967, sect.8-3), and δw is the difference between the gauge value W_0 and the value of the reference potential U on surface S_o. Condition (12.12) imposed on T at infinity follows from (12.6) and the fact that low-degree harmonics (up to degree j_{ref}) of the gravitational potential T are assumed to be known. In contrast to the problem (12.2)–(12.6), S_c is now known and fixed. Thus, by linearization, the problem is reduced to determining the anomalous gravitational potential T outside the surface S only.

In the literature, the problem described by eqns.(12.9)–(12.12) has been investigated as a continuous problem. Let us briefly recall the basic statements proved about the stability of its solution. Holota (1983a,b), Svensson (1983, 1988) and Sansò (1993) proved that the linearized altimetry-gravimetry problem (12.9)–(12.12) is well-posed if $j_{ref} \geq 1$, the anomalous gravitational potential T is taken from the Sobolev space $H_{1/2}(S)$, and boundary functions δw and Δg belong to the Sobolev spaces $H_{1/2}(S_o)$ and $H_{-1/2}(S_c)$, respectively. Being well-posed, in Hadamard's sense, means that the solution exists, and that it is unique and stable. The last property, stability, means that the solution depends continuously on the boundary data. In our particular case, the stability of the solution can be expressed by the following inequality:

$$\| T \|_{H_{1/2}(S)} \leq c \left(\| \delta w \|_{H_{1/2}(S_o)} + \| \Delta g \|_{H_{-1/2}(S_c)} \right) , \qquad (12.13)$$

where c does not depend on δw and Δg.

Svensson (1983, 1988) concluded that condition (12.13) on the stability of the solution implies that the usual least-squares principle

$$\|\delta w - T\|^2_{H_0(S_o)} + \left\|\Delta g - \left(-\frac{\partial T}{\partial r} - \frac{2}{r}T\right)\right\|^2_{H_0(S_c)} \to \min \qquad (12.14)$$

leads to numerical instabilities. On the other hand, on minimizing the Sobolev norms,

$$\|\delta w - T\|^2_{H_{1/2}(S_o)} + \left\|\Delta g - \left(-\frac{\partial T}{\partial r} - \frac{2}{r}T\right)\right\|^2_{H_{-1/2}(S_c)} \to \min , \qquad (12.15)$$

the solution is stable due to condition (12.13).

12.2 Parametrization and discretization

Since we are interested in a global gravity field, the anomalous gravitational potential T will be represented by a solid spherical harmonic series,

$$T(r,\Omega) = \sum_{j=j_{ref}+1}^{\infty} \left(\frac{a}{r}\right)^{j+1} \sum_{m=-j}^{j} T_{jm}Y_{jm}(\Omega) . \qquad (12.16)$$

This parametrization of T ensures that eqns.(12.9) and (12.12) are satisfied in an identical way.

Up to now, the altimetry-gravimetry problem has been formulated in a continuous way. This means that an infinite number of potential coefficients T_{jm} should be determined from boundary conditions (12.10) and (12.11) given continuously at all points on surfaces S_o and S_c. But, in practice, the boundary functionals δw and Δg are measured at discrete points only, and hence the number of potential coefficients T_{jm} that can be determined from such discrete data is finite. The harmonic series (12.16) for T must therefore be truncated at some cut-off degree j_{max}:

$$T(r,\Omega) = \sum_{j=j_{ref}+1}^{j_{max}} \left(\frac{a}{r}\right)^{j+1} \sum_{m=-j}^{j} T_{jm}Y_{jm}(\Omega) . \qquad (12.17)$$

Truncation of the potential series at a finite cut-off degree represents a certain type of regularization, because a high-frequency part of the potential T is excluded from the solution. Let us consider now that the solution of the mixed boundary-value problem (12.9)–(12.12) is searched by the classical least-squares method, i.e., by minimizing the L_2 norms rather than by minimizing the Sobolev norms $\| \delta w \|_{H_{1/2}(S_o)}$ and $\| \Delta g \|_{H_{-1/2}(S_c)}$. The question arises whether the regularization of the solution by truncation of its high-frequency part is efficient enough to stabilize the classical least-squares. In other words: up to which cut-off degree j_{max} is the classical least-squares estimation still numerically stable?

To answer this question, let us formulate the altimetry-gravimetry problem in a discrete form and find a least-squares solution. Let the observations of boundary functional δw and Δg result in finite sets of discrete values δw_i, $i = 1, 2, ..., N_{\delta w}$, and Δg_i, $i = 1, 2, ..., N_{\Delta g}$. Let these data sets create column vectors δw and Δg. Let associated observational errors $e_{\delta w}$ and $e_{\Delta g}$ be governed by the stochastic models of the forms

$$E(e_{\delta w}) = 0 , \qquad E(e_{\Delta g}) = 0 , \qquad (12.18)$$

$$E(e_{\delta w}e_{\delta w}^T) = C_{\delta w} , \qquad E(e_{\Delta g}e_{\Delta g}^T) = C_{\Delta g} , \qquad (12.19)$$

where the symbol E represents the stochastic expectation, and $C_{\delta w}$ and $C_{\Delta g}$ stand for the covarince matrices of the stochastic variables δw and Δg, respectively.

Employing the parametrization (12.17) of potential T, a discrete, stochastic form of boundary conditions (12.10) and (12.11) reads

$$\begin{pmatrix} A_{\delta w} \\ A_{\Delta g} \end{pmatrix} m = \begin{pmatrix} \delta w \\ a\,\Delta g \end{pmatrix} + \begin{pmatrix} e_{\delta w} \\ a\,e_{\Delta g} \end{pmatrix} , \qquad (12.20)$$

where m is a column vector composed of potential coefficients T_{jm}, i.e.,

$$m := \{T_{jm} | j = j_{ref} + 1, \ldots, j_{max}, m = -j, -j+1, \ldots, j\} , \qquad (12.21)$$

and the matrices $A_{\delta w}$ and $A_{\Delta g}$ are formed from a parametrized left-hand side of boundary conditions (12.10) and (12.11):

$$(A_{\delta w})_{i,jm} := \left(\frac{a}{r_i}\right)^{j+1} Y_{jm}(\Omega_i) , \qquad i = 1, 2, \ldots, N_{\delta w} , \quad (12.22)$$

$$(A_{\Delta g})_{i,jm} := (j-1)\left(\frac{a}{r_i}\right)^{j+2} Y_{jm}(\Omega_i) , \qquad i = 1, 2, \ldots, N_{\Delta g} . \quad (12.23)$$

12.3   A least-squares estimation

The solution to the problem formulated above must be searched in an approximate sense because of the existence of observational errors. Let us try to apply the classical discrete least-squares estimator to eqn.(12.20). Let a number of observations δw and Δg be greater than a number of unknown potential coefficients T_{jm}, i.e., $N_{\delta w} + N_{\Delta g} > (j_{max} + 1)^2 - (j_{ref} + 1)^2$. Introducing the following vectors and matrices

$$A := \begin{pmatrix} A_{\delta w} \\ A_{\Delta g} \end{pmatrix} , \qquad d := \begin{pmatrix} \delta w \\ a\,\Delta g \end{pmatrix} , \qquad C := \begin{pmatrix} C_{\delta w} & 0 \\ 0 & a^2 C_{\Delta g} \end{pmatrix} , \quad (12.24)$$

and

$$B := C^{-\frac{1}{2}}A , \qquad y := C^{-\frac{1}{2}}d , \qquad (12.25)$$

the classical least-squares estimate $\hat{m}$ of potential coefficients is given by the system of normal equations (e.g., Tarantola (1987))

$$B^{\dagger}B\hat{m} = B^{\dagger}y \, , \tag{12.26}$$

where the superscript $\dagger$ denotes Hermite conjugation.

In the following numerical examples, we will use simple stochastic models of observation errors assuming that the covariance matrices $C_{\delta w}$ and $C_{\Delta g}$ are proportional to the unit matrix, i.e.,

$$C = \begin{pmatrix} \sigma_{\delta w}^2 I & 0 \\ 0 & a^2 \sigma_{\Delta g}^2 I \end{pmatrix} \, , \tag{12.27}$$

where I is the identity matrix, and $\sigma_{\delta w}$ and $\sigma_{\Delta g}$ are the respective standard deviations of the boundary data δw and Δg. Considering a special form (eqn.(12.27)) of C, the matrix B and the vector y read

$$B = \frac{1}{\sigma_{\delta w}}A_{\delta w} + \frac{1}{a\,\sigma_{\Delta g}}A_{\Delta g} \, , \qquad y = \frac{1}{\sigma_{\delta w}}\delta w + \frac{1}{\sigma_{\Delta g}}\Delta g \, . \tag{12.28}$$

12.4 The axisymmetric geometry

Numerical results computed for an axisymmetric case, when the anomalous potential T as well as the bounding surface S are rotationally symmetric, illustrate the stability property and the approximation error of the altimetry-gravimetry problem. The axisymmetric geometry of S was modelled by zonal spherical harmonics of the global digital terrain model TUG87 (Wieser, 1987) cut at degree 180, the maximum degree of the TUG87 model. Continents were represented by a spherical cap surrounding the north pole with a radius ϑ_0.

The elements of the design matrix B were generated in an equal angular grid with a constant separation $\Delta\vartheta$ between parallels, $\Delta\vartheta = \pi/N$, where N is the number of grid points ($N = N_{\delta w} + N_{\Delta g}$). For such a regular grid, the Nyquist frequency is equal to $\pi/\Delta\vartheta$, and the maximum spherical degree j_{max} up to which the potential coefficients are fully recoverable and are free of aliasing effects is given by the inequality $j_{max} < \pi/\Delta\vartheta$, or $j_{max} < N$. If it is not introduced elsewhere, we will take $N = j_{max} + 1$.

Since we are interested in the question of how the distribution of continents and oceans and the truncation of the potential series influence the stability of the least-squares solution, in the following numerical examples, we will assume that the covariance matrix of boundary data is the diagonal matrix of the form (12.27). As discussed in the introduction, the rms error of the altimeter measurements is about 0.1 m. The standard deviation $\sigma_{\delta w}$ of the anomalous gravity potential data

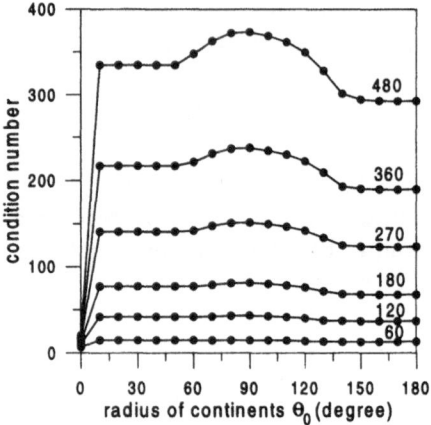

Figure 12.1: Condition number κ of the design matrix B versus the radius of the continental cap ϑ_0 for various cut-off degrees $j_{max} = 60, 120, 180, 270, 360$ and 480 ($j_{ref} = 20$).

δw is then approximately equal to 1 m²s⁻². The rms error of land-borne gravity anomalies used to compute the geoidal heights may be estimated to be 0.3 mGal (Véronneau, 1994, personal communication). In summary, we will use the following estimates of the standard deviations:

$$\sigma_{\delta w} = 1 \text{ m}^2\text{s}^{-2} , \qquad a\,\sigma_{\Delta g} = 20 \text{ m}^2\text{s}^{-2} . \qquad (12.29)$$

In Figure 12.1, we change the radius ϑ_0 of a continental cap from $0°$ to $180°$ and plot the corresponding condition number $\kappa(B)$ of matrix B. Note that the condition number is defined as the ratio of the largest singular value to the smallest singular value (Wilkinson, 1965, sect.2.30). The reference degree of the anomalous potential T is $j_{ref} = 20$, and the maximum degree takes six values, $j_{max} = 60, 120, 180, 270, 360$ and 480. Figure 12.1 shows that the condition number increases with increasing maximum cut-off degree j_{max}. The larger the j_{max}, the worse the numerical conditionality of the matrix B.

The condition number κ takes the minimum values for $\vartheta_0 = 0°$ and $\vartheta_0 = 180°$. Let us discuss these limits. The case $\vartheta_0 = 0°$ corresponds to the external Dirichlet boundary-value problem for the Laplace equation. The fact that κ reaches a minimum for $\vartheta_0 = 0°$ confirms the well-known fact that the Dirichlet problem for an elliptic partial differential equation is stable when changes to the boundary condition are evaluated in the L_2 norm (Rektorys, 1977).

The other limit, $\vartheta_0 = 180°$, represents the case when the solution to the Laplace equation is only constrained by the boundary condition (12.11) prescribed on the whole surface S. The fact that κ takes the second minimum for this case is in full agreement with theoretical results on the stability of the altimetry-gravimetry problem of the second kind (Holota, 1983a,b; Sacerdotte

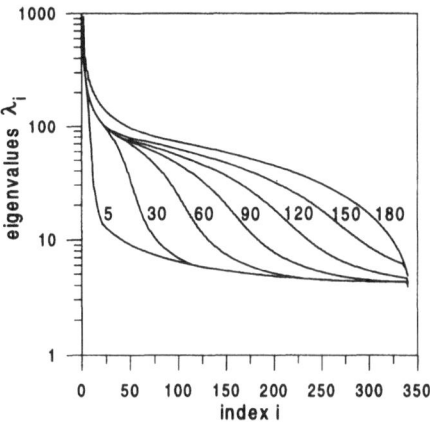

Figure 12.2: Spectra of the design matrix B for various radii of the continental cap $\vartheta_0 = 5°$, $30°$, $60°$, $90°$, $120°$, $150°$ and $180°$ ($j_{ref} = 20$ and $j_{max} = 360$).

and Sansò, 1983). Svensson (1988) and Sansò (1993) proved that the altimetry-gravimetry problem of the second kind has a unique solution (in Hadamard's sense) if gravity anomaly Δg and gravity disturbance δg are square integrable on respective surfaces S_c and S_o, and the anomalous potential T is a function of the Sobolev space $H_1(S)$. As a consequence, the minimization of the L_2-norms $\| \Delta g \|_{L_2(S_c)}$ and $\| \delta g \|_{L_2(S_o)}$ results in a unique and stable solution.

Inspecting Figure 12.1, we can observe that, for a given j_{max} and $\vartheta_0 > 0°$, the condition number κ does not vary significantly with the cap radius ϑ_0. For example, in the worst case $j_{max} = 480$, the ratio $\kappa_{\vartheta_0=90°}$ to $\kappa_{\vartheta_0=180°}$ is equal to 1.3. Moreover, the ratio $\kappa_{\vartheta_0=90°}$ to $\kappa_{\vartheta_0=0°}$ is equal to 18.3 only. Judging from the magnitude of κ only, the conditionality of the matrix B for the mixed problem is comparable to that with $\vartheta_0 = 180°$ and is slightly worse than in the case of $\vartheta_0 = 0°$. Since $1/\kappa_{\vartheta_0=90°,j_{max}=480}$ ($\doteq 1/370$) is much greater than the machine floating point precision ε, we can conclude that the matrix B is well-conditioned for any radius ϑ_0 (and, of course, when $j_{max} \leq 480$).

Figure 12.2 yields an alternative view to this result. It plots the sizes of singular values λ_i against their index number (i.e., the spectrum of B) depending upon the radius ϑ_0 of the continental cap. As already shown in Figure 12.1, the sizes of the maximum and minimum singular values, λ_{max} and λ_{max}, only weakly depend on ϑ_0 ($\vartheta_0 > 0°$). On the other hand, the shape of the spectrum of matrix B depends strongly on the size of ϑ_0. For a small value ϑ_0, but $\vartheta_0 > 0°$, the spectrum is spiky and resembles the delta function, i.e., there are a number of small λ_i's in the spectrum of B. As the size of the continental cap increases, the number of small λ_i's decreases and the similarity of the spectrum to the delta function disappears.

Let us investigate the influence of the number N of data grid points, $N = N_{\delta w} + N_{\Delta g}$, on the conditionality of the design matrix B. To do this, we choose

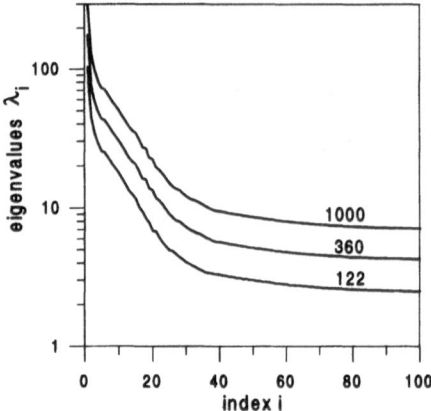

Figure 12.3: Spectra of the design matrix B for different numbers of data points, $N = 122$, 360 and 1000 ($j_{ref} = 20$ and $j_{max} = 60$). The condition number κ is the same for all three cases, $\kappa = 41.7$.

the fixed reference and maximum spherical degrees, $j_{ref} = 20$ and $j_{max} = 120$, the radius of the continental cap, $\vartheta_0 = 30°$, and we increase the number of data points from $N = 122$ to $N = 1000$. Figure 12.3 shows the spectra of matrix B for three different values of N: $N = 122$, 360, and 1000. As a result, the condition number κ does not depend on the number of data points N; in our example, $\kappa = 41.7$ for all N values. Similarly, apart from a constant shift, the shape of the spectrum of B is practically independent of the number of grid points.

Now, let us turn our attention to the approximation error of the least-squares solution. The approximation error will be estimated by the L_2-norm of the least

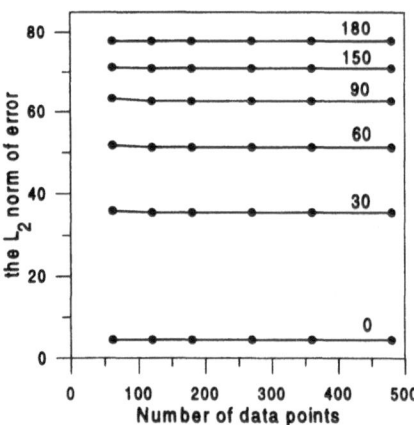

Figure 12.4: The square root of the L_2-norm of the least favourable error ϵ_ℓ against the number of data points N for continental caps of different sizes $\vartheta_0 = 0°$, $30°$, $60°$, $90°$, $150°$ and $180°$ ($j_{ref} = 20$ and $j_{max} = 60$).

favourable error ϵ_ℓ (Svensson, 1988),

$$\| \epsilon_\ell \|_{L_2}^2 \approx s_\ell \left(N_{\delta w} \sigma_{\delta w}^2 + N_{\Delta g} a^2 \sigma_{\Delta g}^2 \right) . \tag{12.30}$$

This relation holds under the assumptions that $N_{\delta w} + N_{\Delta g} \to \infty$, that observations are independent, i.e., their covariance matrix has the form of eqn.(12.27), and that s_ℓ is the largest eigenvalue of the matrix $(\boldsymbol{B}^\dagger \boldsymbol{B})^{-1}$. Considering singular value decomposition of matrix $\boldsymbol{B}$ with singular values λ_i, s_ℓ is equal to the reciprocal value of λ_{min}^2, i.e., $s_\ell = 1/\lambda_{min}^2$.

We choose the fixed reference and maximum spherical degrees, $j_{ref} = 20$ and $j_{max} = 60$, and we change the radius ϑ_0 of the continental cap from $\vartheta_0 = 0°$ to $\vartheta_0 = 180°$ and the number N of data points from $N = 62$ to $N = 480$. Figure 12.4 plots the square root of the L_2-norm of the least favourable error ϵ_ℓ. We can observe that $\left(\| \epsilon_\ell \|_{L_2}^2 \right)^{1/2}$ increases in a non-linear way with an increase in the size of the cap radius ϑ_0, and that it is independent of the number N of observations. The latter fact strongly contradicts Svensson's numerical example (Svensson, 1988, Fig.7), showing that the the L_2-norm of the least favourable error ϵ_ℓ of the least-squares solution of the mixed boundary-value problem grows with an increase in the number of data points. Unfortunately, his synthetic example cannot be recomputed because information about the cut-off degrees j_{ref} and j_{max} is missing. Moreover, he did not compute $\| \epsilon_\ell \|_{L_2}^2$ according to the approximate formula (12.30) but only estimated its magnitude by computing the trace of matrix $\boldsymbol{B}(\boldsymbol{B}^\dagger \boldsymbol{B})^{-2} \boldsymbol{B}^\dagger$. He also used a very small, and hence unrealistic, estimate for the standard deviation of gravity anomalies, $\sigma_{\Delta g} = 16\mu$Gal. Such accurate gravity data are still not available over all the continents for solving the global altimetry-gravimetry problem. Since Svensson's numerical results can be neither proved nor disproved, we will follow our numerical test and conclude that L_2-norm of the least favourable approximation error does not depend on the number of data points, but only on the size of the continental cap.

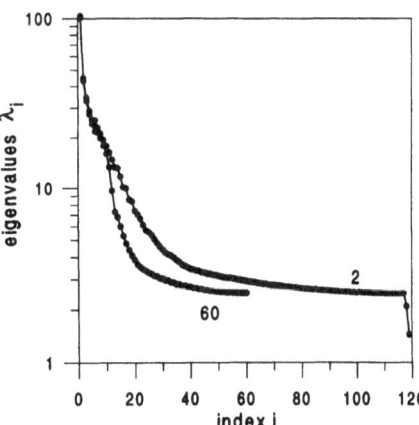

Figure 12.5: Spectra of the design matrix $\boldsymbol{B}$ for different reference degrees $j_{ref} = 2$ and 60 ($j_{max} = 120$ and $\vartheta = 30°$).

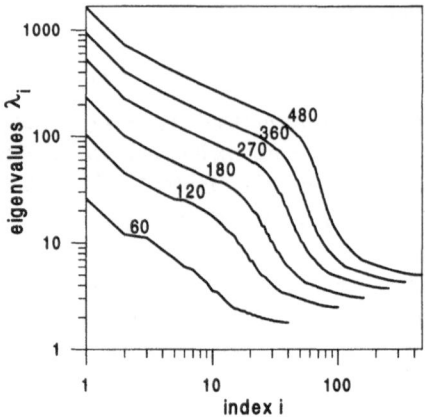

Figure 12.6: Spectra of the design matrix B for various cut-off degrees $j_{max} = 60$ ($\kappa = 14$), 120 (41), 180 (76), 270 (140), 360 (217) and 480 (334) ($j_{ref} = 20$ and $\vartheta_0 = 30°$).

Another question that arises is how does a decrease in the number of low-degree harmonics of the anomalous potential T affect the conditionality of matrix B. We have computed the spectra of B for two cases: $j_{ref} = 2$ and $j_{ref} = 60$. The maximum degree j_{max} was equal to 120, and the radius ϑ_0 of the continental cap was 30°. The result, shown in Figure 12.5, indicates that reducing the low-degree harmonics of potential T causes a decrease in the smallest singular values λ_i's of matrix B. For the example plotted in Figure 12.5, the condition number κ is reduced from 71.9 to 40.7, when the first 60 harmonics are left from the anomalous potential T. From a numerical point of view, the design matrix B becomes more stable.

A more pronounced role on the conditionality of matrix B comes into play when a high frequency part of potential T is cut. To provide a quantitative view of this effect, we have plotted the spectra of matrix B for the case when the continental cap has radius $\vartheta_0 = 30°$, the reference degree $j_{ref} = 20$, and maximum degree j_{max} of the potential series (12.17) takes the values 60, 120, 180, 270, 360 and 480. From Figure 12.6 we can observe that the condition number of matrix B increases with an increase in the size of the cut-off degree j_{max}, so that the matrix B turns out to be less stable for a larger cut-off degree j_{max}. This fully agrees with the instability property of the least-squares solution of a continuous mixed problem for which $j_{max} \to \infty$. Nevertheless, for the example plotted in Figure 12.6, $\kappa = 334.4$ in the worst case with $j_{max} = 480$. Realizing that 1/334.4 is much greater than a machine's floating point precision – even for single-precision arithmetic – the matrix B still may be considered as well-conditioned for that case.

In other words, choosing a priori a certain maximum degree of potential series (12.17), we truncate high-frequency components of the gravitational potential and somehow stabilize the solution of the mixed boundary-value problem. Figure 12.6

shows that this type of regularization is efficient enough, at least up to degree $j_{max} = 480$ for a continental cap of radius $\vartheta_0 = 30°$. When the potential series is cut at a higher degree, the matrix $\boldsymbol{B}$ is closer to an ill-conditioned one.

12.5 Overdetermination

In this section, we shall investigate another type of stabilization of the least-squares solution, consisting of additional information on the gravity field. In accordance with reality, we will assume that the sea surface is partly covered by gravity measurements obtained using ships. When this extra information is supplied, the altimetry-gravimetry problems becomes overdetermined.

Subtracting the reference gravity γ and the gravitational effect of the sea surface topography from the sea surface gravity measurements, we obtain data on gravity disturbance δg. We will assume that the gravity disturbance data are known on S_d, a part of the sea surface S_o, whereas potential disturbance data are still known over the whole surface S_c. The overdetermined altimetry-gravimetry problem can then be described by the following equations:

$$\nabla^2 T = 0 \qquad \text{outside } S \ , \tag{12.31}$$

$$T = \delta w \qquad \text{on } S_o \ , \tag{12.32}$$

$$\frac{\partial T}{\partial r} = \delta g \qquad \text{on } S_d \ , \tag{12.33}$$

$$-\frac{\partial T}{\partial r} - \frac{2}{r} T = \Delta g \qquad \text{on } S_c \ , \tag{12.34}$$

$$T \sim O\left(\frac{1}{r^{j_{ref}+2}}\right) \qquad r \to \infty \ , \tag{12.35}$$

where

$$S = S_o \cup S_c \ , \tag{12.36}$$

$$S_d \cap S_o \neq 0 \ , \tag{12.37}$$

$$S_d \cap S_c = \emptyset \ , \tag{12.38}$$

$$\text{meas } S_d > 0 \ . \tag{12.39}$$

To discretize the problem (12.31)–(12.35), we will proceed in the same way as in the previous sections. Let observations of the gravity disturbance δg result in a set of discrete values δg_i, $i = 1, 2, ..., N_{\delta g}$. Let δg be a column vector created from data δg_i. If we use parametrization (12.17) of anomalous potential T, the discrete forms of boundary conditions (12.32)–(12.34) are as follows:

$$\begin{pmatrix} \boldsymbol{A}_{\delta w} \\ \boldsymbol{A}_{\Delta g} \\ \boldsymbol{A}_{\delta g} \end{pmatrix} \boldsymbol{m} = \begin{pmatrix} \delta w \\ a\,\Delta g \\ a\,\delta g \end{pmatrix} + \begin{pmatrix} \boldsymbol{e}_{\delta w} \\ a\,\boldsymbol{e}_{\Delta g} \\ a\,\boldsymbol{e}_{\delta g} \end{pmatrix} \ , \tag{12.40}$$

where matrices $\boldsymbol{A}_{\delta w}$ and $\boldsymbol{A}_{\Delta g}$ are given by eqns.(12.22) and (12.23), and

$$(\boldsymbol{A}_{\delta g})_{i,jm} := -(j+1)\left(\frac{a}{r_i}\right)^{j+2} Y_{jm}(\Omega_i) \ , \qquad i = 1, 2, \ldots, N_{\delta g} \ . \tag{12.41}$$

The least-squares solution $\hat{m}$ to problem (12.40) can again be expressed by normal equations (12.26) with matrices extended for the gravity disturbance data:

$$A := \begin{pmatrix} A_{\delta w} \\ A_{\Delta g} \\ A_{\delta g} \end{pmatrix} \;, \qquad d := \begin{pmatrix} \delta w \\ a\,\Delta g \\ a\,\delta g \end{pmatrix} \;, \qquad C := \begin{pmatrix} C_{\delta w} & 0 & 0 \\ 0 & a^2 C_{\Delta g} & 0 \\ 0 & 0 & a^2 C_{\delta g} \end{pmatrix} \;.$$
(12.42)

In a particular case, when the covariance matrix C has a diagonal form

$$C = \begin{pmatrix} \sigma_{\delta w}^2 I & 0 & 0 \\ 0 & a^2 \sigma_{\Delta g}^2 I & 0 \\ 0 & 0 & a^2 \sigma_{\delta g}^2 I \end{pmatrix} \;,$$
(12.43)

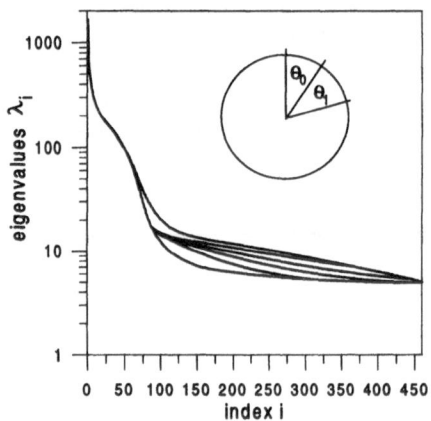

Figure 12.7: Spectra of the design matrix B for various radii ϑ_1 of the overlapping surface S_d, $\vartheta_1 = 0°$, $30°$, $60°$, $90°$, $120°$ and $150°$ ($\vartheta_0 = 30°$, $j_{ref} = 20$ and $j_{max} = 480$).

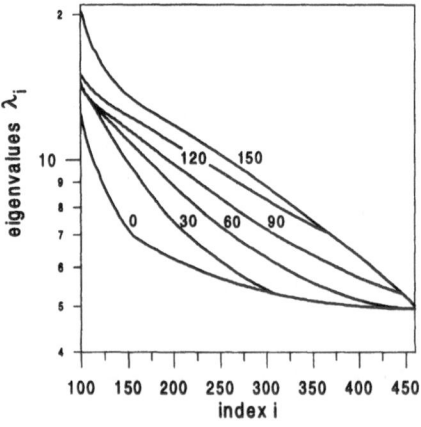

Figure 12.8: A detail of Fig. 7.

where $\sigma_{\delta g}$ is the standard deviation of the gravity disturbance data δg_i, the design matrix $\boldsymbol{B}$ and vector $\boldsymbol{y}$ may be written as

$$\boldsymbol{B} := \frac{1}{\sigma_{\delta w}} \boldsymbol{A}_{\delta w} + \frac{1}{a\,\sigma_{\Delta g}} \boldsymbol{A}_{\Delta g} + \frac{1}{a\,\sigma_{\delta g}} \boldsymbol{A}_{\delta g} , \qquad \boldsymbol{y} := \frac{1}{\sigma_{\delta w}} \delta\boldsymbol{w} + \frac{1}{\sigma_{\Delta g}} \Delta\boldsymbol{g} + \frac{1}{\sigma_{\delta g}} \delta\boldsymbol{g} .$$
$$(12.44)$$

Matrix (12.44) differs from matrix (12.28) by $\boldsymbol{A}_{\delta g}/a\,\sigma_{\delta g}$. Examining the structures of matrices $\boldsymbol{A}_{\delta w}$, $\boldsymbol{A}_{\Delta g}$ and $\boldsymbol{A}_{\delta g}$, we find that matrix $\boldsymbol{A}_{\delta g}^{\dagger}\boldsymbol{A}_{\delta g}/a^2\sigma_{\delta g}^2$, occurring in the matrix $\boldsymbol{B}^{\dagger}\boldsymbol{B}$ of normal equations for the overdetermined case, regularizes the matrix $\boldsymbol{B}^{\dagger}\boldsymbol{B}$ in a certain manner. Hence, the corresponding least-squares solution may be classified as the Levenberg-Marquardt (Levenberg, 1944; Marquardt, 1963) or damped least-squares solution. The following numerical examples yield the quantitative view of the regularization.

12.6   Numerical examples

Axisymmetric geometry

We will begin by presenting numerical results on the stabilization of the overdetermined mixed boundary-value problem for the case with axisymmetric geometry. As in the previous numerical examples, continents are represented by a spherical cap of radius ϑ_0 surrounding the north pole. The surface S_d, where both the gravity potential disturbance δw and the gravity disturbance δg are prescribed, is represented by a spherical ring of radius ϑ_1 (for a schematic sketch see Figure 12.7). The covariance matrices of the boundary data will again be approximated by the diagonal matrices with variances $\sigma_{\delta w}^2$, $\sigma_{\Delta g}^2$ and $\sigma_{\delta g}^2$ appearing in the diagonals. Numerical values of $\sigma_{\delta w}$ and $\sigma_{\Delta g}$ are taken according to eqn.(12.29). The rms error of sea-borne gravity disturbances may be estimated as 3 mGal (Véronneau, 1994, personal communication), i.e., the value

$$a\,\sigma_{\delta g} = 200 \text{ m}^2\text{s}^{-2} , \qquad (12.45)$$

will be used in the following examples.

Figure 12.7, together with a detailed view in Figure 12.8, plot the spectrum of matrix $\boldsymbol{B}$, defined by eqn.(12.44), for the continental cap of radius $\vartheta_0 = 30°$; the radius ϑ_1 of overlapping surface S_d takes six values from $\vartheta_1 = 0°$ to $\vartheta_1 = 150°$. The condition number is approximately the same for all the cases, $\kappa \doteq 330$, i.e., the matrix $\boldsymbol{B}$ is always well-conditioned. We can see that the effect of overdetermination on the shape of the spectrum of $\boldsymbol{B}$ is not too large. This is because the relative accuracy of the sea-borne gravity observations is much less than the accuracy of the altimeter measurements and land-borne gravity observations. Consequently, the parameter $1/a^2\sigma_{\delta g}^2$ appearing in the 'regularizing' matrix $\boldsymbol{A}_{\delta g}^{\dagger}\boldsymbol{A}_{\delta g}$ in the matrix of normal equations is small compared with factors $1/\sigma_{\delta w}^2$ and $1/a^2\sigma_{\Delta g}^2$, and thus the contribution of the term $\boldsymbol{A}_{\delta g}^{\dagger}\boldsymbol{A}_{\delta g}/a^2\sigma_{\delta g}^2$ to the matrix of normal equations is not very significant.

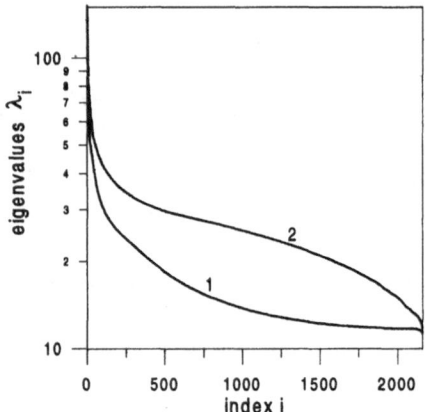

Figure 12.9: Spectra of the design matrix B for the actual geometry of the Earth's surface and various sizes of the overlapping surface S_d ($j_{ref} = 20$ and $j_{max} = 50$). Curve '1' stands for all the spectra with overlapping $S_d = 0$, $S_d = \frac{1}{8}S_o$, $S_d = \frac{1}{4}S_o$, $S_d = \frac{1}{2}S_o$, and $S_d = \frac{3}{4}S_o$, whereas curve '2' shows the spectrum with overlapping $S_d = S_o$.

Actual geometry

The geometrical property of the bounding surface S is now described by zonal as well as tesseral and sectoral spherical harmonics of the global digital terrain model TUG87 (Wieser, 1987) cut at degree 180. Figure 12.9 plots the spectrum of the design matrix B for the case when the reference and maximum spherical degrees of the potential series (12.17) are chosen as $j_{ref} = 20$ and $j_{max} = 50$. The shape and size of the continental surface S_c have been computed from the TUG87 model, so that S_c approaches the actual distribution of the continents. The overlapping surface S_d is adjacent to the continents; its size increases from zero to S_o. The last limit corresponds to the case when the whole sea surface S_o is covered by both altimeter and gravity measurements. The spectra of the design matrix B for the overlapping area S_d of sizes from $S_d = 0$ to $S_d = \frac{3}{4}S_o$ differ only slightly, so that they cannot be distinguished in Figure 12.9; the curve '1' plots the spectra of matrix B for $S_d = 0$, $S_d = \frac{1}{8}S_o$, $S_d = \frac{1}{4}S_o$, $S_d = \frac{1}{2}S_o$, and $S_d = \frac{3}{4}S_o$. The insensitivity of the spectrum of B to the overdetermination caused by extra gravity data is again a consequence of the fact that the relative accuracy of sea-borne gravity observations is much lower than the accuracy of the altimeter measurements and land-borne gravity observations.

12.7 Conclusion

It is a well-known fact that the least-squares solution to the *continuous* altimetry-gravimetry problem displays an instability behaviour. This chapter was motivated by the question of how much the least-squares solution to the *discrete* altimetry-gravimetry problem for a global gravity field determination reflects this undesirable property. In other words, how can the least-squares solution of the discrete altimetry-gravimetry problem be regularized. We have investigated two possible ways to achieve this regularization. First, truncating a spherical harmonics series of the anomalous potential at a finite cut-off degree excludes a high frequency part of the solution, and thus regularizes the solution. The efficiency of this regularization is strongly dependent on the ratio of the relative accuracies of altimetry and gravity data. Considering that the relative accuracy of altimetry observations is about ten times better than that of gravity measurements – which approximately holds for the currently available data – we have demonstrated that this kind of regularization is efficient enough when a global gravity model is constructed up to spherical degrees 500.

We have also treated another way of stabilization of the mixed boundary-value problem. This consists of supplying the boundary data with gravity disturbances over the oceans. Overdetermining with extra boundary data represents a natural method of stabilization; namely, it increases the magnitudes of the smallest singular values of the design matrix. The corresponding least-squares solution, which may be classified as the Levenberg-Marquardt or damped least-squares solution, is then more stable than the solution of the original altimetry-gravimetry problem. As in the previous case, the efficiency of such a regularization strongly depends on the accuracy of the sea-borne gravity data. Since the accuracy of today's ship gravity observations is about 10 times worse than that of land gravity observations, the effect of stabilization of the least-squares solution by overdetermination is not very significant.

Svensson (1988) has argued that even the least-squares solution is stable, its approximation error grows with an increase in the number of boundary data points. Keeping this fact in mind, we computed the approximation error, or, its estimate in the least favourable case, for a model with axisymmetric geometry. Our numerical example showed that, for the problem studied here, the approximation error of the least-squares solution was only sensitive to the size of the continental area and not to the number of data points. Since this last conclusion contradicts Svensson's numerical experience, we tried to recompute Svensson's synthetic example. We did not succeed due to lack of information about Svensson's example.

In conclusion, we have demonstrated that the classical least-squares approach does not fail in the case when the discrete altimetry-gravimetry boundary-value problem is to be solved for global gravity field determination. The approach enables us to combine physically different types of boundary data with different relative accuracies. The stabilization of the problem is mainly achieved by cutting a high-frequency part of the gravity spectrum. Another important property of the method suggested here is that it avoids having to convert altimeter data to gravity disturbances. Despite the instability property of such a transformation, it is often used as the preparatory step in constructing recent global gravity models.

Summary

Let us briefly summarize main results of this work.

- One of the main conclusions is that the solution to the Stokes two-boundary-value problem cannot be guaranteed for all wavelengths. The spectral analysis of the boundary operator for harmonic functions regular at infinity has revealed that the eigenvalue spectrum may contain a null eigenvalue or an eigenvalue of a very small size. Consequently, the inverse operator is ill-posed or even does not exist. Once this ill-posed case occurs, the Stokes two-boundary-value problem must be regularized such that this null eigenvalue and its vicinity are excluded from the solution.

- For a mountainous terrain, such as the Rocky Mountains, it is not sufficient to model then topographical density by a constant value ϱ_0 equal to a mean crustal density 2.67 g/cm^3 when a 'decimetre geoid' is to be compiled. The topographical density corrections due to lateral density inhomogeneities may reach several decimetres, and, hence, they must be considered in the final computation. On the other hand, in a flat low-land terrain, the approximation of the actual topographical density by ϱ_0 is acceptable for a 'decimetre geoid' computation.

- Although it is a common belief that there is not significant for 'decimetre geoid' whether spherical or planar approximation of the geoid is used to compute the topographical effects, we beg to differ. The type of approximations significantly influences the Bouguer parts of topographical terms; the bias due to planar approximation of the geoid may reach a few metres.

- The Taylor series expansion of the Newton kernel does not converge in a rugged mountainous terrain such as the Rocky Mountains. The divergency of the Taylor series may cause errors of the order of several decimetres in terms of geoidal heights. The formulae derived herein overcome the problem of divergency of the Taylor series; they may be used to evaluate the gravitational effect of a very rugged terrain.

- We recommend to carry out the spectral analysis of gravity anomalies for different compensation/condensation models before geoid height computation and to choose that compensation model which reduces a high frequency part of surface gravity anomalies in a most efficient way. The experience with the spectral analysis of gravity anomalies from the region of the Canadian Rocky Mountains indicates that rather the Airy-Heiskanen model than Helmert's 2nd condensation technique should be used to reduce high-frequency oscillations of surface gravity data.

- There is a number of possibilities how to take into account a satellite gravitational model as a reference in the Stokes two-boundary-value problem. Herein, we have formulated the Stokes two-boundary-value problem for a high-frequency part of the anomalous gravitational potential such that low-frequency reference potential does not depend on the way of compensation of topographical masses but only on a satellite gravitational model and the gravitational field generated by topographical masses.

- The discrete downward continuation problem for geoid determination is undoubtedly stable and its solution may be carry out by Jacobi's iterative method once the grid step size of the surface observations as well as of the discrete solution is not smaller than 5 arcmin. This conclusion drawn for the Canadian Rocky Mountains will be valid anywhere else in the world, with perhaps the exception of Himalayas, since the Rockies represent one of the highest and roughest mountainous terrain pattern.

- The solution to the Stokes boundary-value problem on an ellipsoid of revolution with the accuracy of the order up $O(e_0^2)$ (i.e., omitting terms of $O(e_0^4)$ and higher) has been found in a closed form. The ellipsoidal Stokes function describing the effect of the ellipticity of boundary on the solution to Stokes's boundary-value problem can be approximated by function $1/\psi$ in a neighbourhood of its singular point $\psi = 0$. Hence, the degree of singularity of the ellipsoidal Stokes function at the point $\psi = 0$ is the same as that of the spherical Stokes function.

- Green's function to the external Dirichlet boundary-value problem for the Laplace equation with data distributed on an ellipsoid of revolution with the accuracy of the order up $O(e_0^2)$ has been constructed in a closed form. The ellipsoidal Poisson kernel describing the effect of the ellipticity of the boundary on the solution to this boundary-value problem has been expressed as a finite sum of elementary functions which describe analytically the behaviour of ellipsoidal Poisson kernel at the singular point $\psi = 0$. We have shown that the degree of singularity of the ellipsoidal Poisson kernel in the vicinity of its singular point is of the same degree as that of the original spherical Poisson kernel.

- Green's function to the boundary-value problem of Stokes's type with ellipsoidal corrections in the boundary condition for anomalous gravity has constructed in a closed form. The 'spherical-ellipsoidal' Stokes function describing the effect of two ellipsoidal correcting terms occurring in the boundary condition for anomalous gravity is expressed in $O(e_0^2)$-approximation as a finite sum of elementary functions analytically representing the behaviour of the integration kernel at the singular point $\psi = 0$. We show that 'spherical-ellipsoidal' Stokes's function has only a logarithmic singularity in the vicinity of its singular point. The constructed Green function enables us to avoid applying an iterative approach to solve Stokes's boundary-value problem with ellipsoidal correction terms involved in boundary condition for anomalous gravity. A new Green function approach is more convenient from numerical point of view since the solution to the boundary-value problem is determined in one step by computing a Stokes-type integral. The question on the convergency of an iterative scheme recommended so far to solve this boundary-value problem is thus irrelevant.

• We demonstrate that the least-squares solution to the discrete altimetry-gravimetry problem is stable provided that the boundary functionals are discretized in an equal angular grid, the cut-off degree of a global gravity model is not greater than 500, and the covariance matrices of the boundary data are set up in accordance with today's accuracy of geodetic measurements. Moreover, provided that the number of observations is greater than the cut-off degree of the potential series, the approximation error of the least-squares solution does not depend on the number of observations. We paper also investigate the numerical stability of the discrete altimetry-gravimetry problem overdetermined by extra gravity data measured by ships over a part of the sea surface. Overlapping boundary data does not have, however, a significant impact on the conditionality of the matrix of normal equations due to the low accuracy of sea-borne gravity observations.

In all of the discussions above, we have not addressed the question of height and gravity data accuracy. The effects of spatial data distribution (irregularity and sparseness) and data accuracy are probably very significant. This also prevented us to investigate the effect of various Stokes's kernel modification schemes which differ in the weighting the near-zone and far-zone contributions to the Stokes integral. Only the stochastic model of surface gravity data errors can be used as an objective function to perform such weighting. There exists surely a number of other questions connected with a precise geoid determination which have not been sorted out in this work. The research reported here should be thus understood as a part of our ultimate goal.

References

Abramowitz, M. and I.A. Stegun (1970). *Handbook of Mathematical Functions*, Dover Publ., New York. [quoted in chapters 9,10]

Arfken, G. (1968). *Mathematical Methods for Physicists*. Academic Press, 2.ed., New York and London. [9]

Bašić, T. and R.H. Rapp (1991). Oceanwide predictions of gravity anomalies and sea surface heights using Geos-3, Seasat and Geosat altimeter data, and ETOPO5U bathymetric data. Dept. of Geod. Sci. Rep., **416**, Ohio State Univ., Columbus. [12]

Bjerhammar, A. (1962). Gravity reduction to a spherical surface. Report of the Royal Institute of Technology, Geodesy Division, Stockholm. [8]

Bjerhammar, A. (1963). A new theory of gravimetric geodesy. Report of the Royal Institute of Technology, Geodesy Division, Stockholm. [8]

Bjerhammar, A. (1976). A Dirac Approach to Physical Geodesy. *Zeitschrift für Vermessungswesen*, **101**, 41-44. [8]

Bjerhammar, A. (1987). Discrete Physical geodesy. Rep. **380**, Dept. of Geodetic Science and Surveying, The Ohio State University, Columbus. [8,12]

Colombo, O.L. (1981). Numerical methods for harmonic analysis on the sphere. Rep. **310**, Dept. of Geodetic Science and Surveying, The Ohio State University, Columbus. [8]

Cruz, J.Y. (1985). Disturbance vector in space from surface gravity anomalies using complementary models. Rep. **366**, Dept. of Geodetic Science and Surveying, The Ohio State University, Columbus. [1,8,11]

Cruz, J.Y. (1986). Ellipsoidal corrections to potential coefficients obtained from gravity anomaly data on the ellipsoid. Rep. **371**, Dept. of Geodetic Science and Surveying, The Ohio State University, Columbus. [1,8]

Denker, H., D. Behrend and W. Torge (1994). European gravimetric geoid: status report 1994. In *Gravity and Geoid*, pp. 423-432, eds. H. Sünkel and I. Marson, IAU Symposium 113, Springer Verlag, New York. [0]

Engels, J., E. Grafarend, W. Keller, Z. Martinec, F. Sansò and P. Vaníček (1993). The geoid as an inverse problem to be regularized. In: *Inverse Problems: Principles and Applications in Geophysics, Technology and Medicine*. Eds. G.Anger, R.Gorenflo, H.Jochmann, H.Moritz and W.Webers, Akademie-Verlag, Berlin, 122-167. [8]

Featherstone, W.E. (1992). A G.P.S. controlled gravimetric determination of the geoid of the British Isles. PH.D. Thesis, University of Oxford, England, pp.252.   [0,6]

Forsberg, R. and M.G. Sideris (1993). Geoid computations by the multi-band spherical FFT aproach. *Man. Geod.*, **18**, 82-90.   [0,5,6]

Geological Map of British Columbia (1962), Map (932), Geological Survey of Canada.   [6]

Geological Map of Canada (1962), Map 1045A, Geological Survey of Canada. [6]

Gradshteyn, I.S. and I.M. Ryzhik (1980). *Tables of Integrals, Series, and Products*. Corrected and enlarged edition, transl. by A.Jeffrey, Academic Press, New York.   [3,9]

Groetsch, C.W. (1984). *The Theory of Tikhonov Regularization for Fredholm equations of the first kind.* Pitman Publishing, Boston.   [8]

Gysen, H. (1994). Thin-plate spline quadrature of geodetic integrals. *Bull. Géod.*, **68**, 173-179.   [8]

Hackbusch, W. (1995). *Integral Equations. Theory and Numerical Treatment.* Birkhäuser Verlag, Basel.   [10]

Hansen, P.C. (1992). Numerical tools for analysis and solution of Fredholm integral equations of the first kind. *Inverse Problems*, **8**, 849-872.   [8]

Heck, B. (1991). On the linearized boundary value problem of physical geodesy. Rep. **407**, Dept. of Geodetic Science and Surveying, The Ohio State University, Columbus.   [7,9,11]

Heck, B. (1993). A revision of Helmert's second method of condensation in geoid and quasigeoid determination. Paper presented at 7th Int. Symposium of Geodesy and Physics of the Earth, IAG Symposia Proceedings, No. 112, Springer 1993, p.246-251.   [1,3,4,5]

Heiskanen, W.H. and H. Moritz (1967). *Physical Geodesy.* W.H.Freeman and Co., San Francisco.   [0,1,2,3,5,6,7,8,9,11,12]

Helmert, F.R. (1884). *Die mathematischen und physikalischen Theorieen der höheren Geodäsie*, vol.2. Leipzig, B.G.Teubner (reprinted in 1962 by Minerva GMBH, Frankfurt/Main).   [1]

Hipkin, R.G. (1994). How close are we to a centimetric geoid? In *Gravity and Geoid*, pp. 529-538, eds. H. Sünkel and I. Marson, IAU Symposium 113, Springer Verlag, New York.   [0]

Hobson, E.W. (1965). *The Theory of Spherical and Ellipsoidal Harmonics.* Chelsea Publ. Comp., 2.ed., New York.   [9]

Holota, P. (1983a). The altimetry-gravimetry boundary value problem I: Linearization, Fridrich's inequality. *Bollettino di Geodesia e Scienze Affini*, **42**, 13-32.   [12]

Holota, P. (1983b). The altimetry-gravimetry boundary value problem II: Weak solution, V-ellipticity. *Bollettino di Geodesia e Scienze Affini*, **42**, 69-84.   [12]

Hörmander, L. (1976). The boundary problems of physical geodesy. *Archive for Rational Mechanics and Analysis*, **62**, 1-52. [1]

Ilk, K.H. (1987). On the regularization of ill-posed problems. In: *Figure and Dynamics of the Earth, Moon, and Planets*, ed. P.Holota, Research Inst. of Geodesy, Topography and Cartography, Prague, 365-383. [8]

Ilk, K.H. (1993). Regularization for high resolution gravity field recovery by future satellite techniques. In: *Inverse Problems: Principles and Applications in Geophysics, Technology and Medicine*. Eds. G. Anger, R. Gorenflo, H. Jochmann, H. Moritz and W. Webers, Akademie-Verlag, Berlin, 189-214. [8]

Isaacson, E. and H.B. Keller (1973). *Analyse Numerischer Verfahren*. Verlag Harri Deutsch, Zürich. [11]

Jekeli, C. (1981). The downward continuation to the Earth's surface of truncated spherical and ellipsoidal harmonic series of the gravity and height anomalies. Rep. **323**, Dept. of Geodetic Science and Surveying, The Ohio State University, Columbus. [1,11]

Jekeli, C. (1988). The exact transformation between ellipsoidal and spherical harmonic expansions. *Man. Geod.*, **13**, 106-113. [7]

Kellogg, O.D. (1929). *Foundations of Potential Theory*. Berlin, J.Springer (reprinted by Dover Publications, New York, 1953). [7,8,10]

Kondo, J. (1991). *Integral Equations*. Clarendon Press, Oxford. [8]

Lapidus, L. and G.F. Pinder (1982). *Numerical Solution of Partial Differential Equations in Science and Engineering*. J.Wiley, New York. [1]

Lerch, F.J., et al., (1992). Geopotential models of the Earth from satellite tracking, altimeter and surface gravity observations: GEM-T3 and GEM-T3S. NASA Technical Memorandum 104555, Goddard Space Flight Center, Greenbelt, MD. [12]

Levenberg, K. (1944). A method for the solution of certain non-linear problems in least squares. *Quant. Appl. Math.*, **2**, 164-168. [12]

Mainville, A., M. Véronneau, R. Forsberg and M. Sideris (1995). A comparison of geoid and quasigeoid modeling methods in rough topography. In *Gravity and Geoid*, pp. 491-501, eds. H. Sünkel and I. Marson, IAU Symposium 113, Springer Verlag, New York. [0]

Mangulis, V. (1965). *Handbook of Series for Scientists and Engineers*. Academic Press, New York. [9,10]

Marple, S.L. (1987). *Digital Spectral Analysis with Applications*. Prentice-Hall, Inc., Englewood Cliffs, New Jersey. [8]

Marquardt, D.W. (1963). An algorithm for least-squares estimation of nonlinear parameters. *J. Soc. Ind. Appl. Math.*, **11**, 431-441. [12]

Martinec, Z. (1989). Program to calculate the spectral harmonic expansion coefficients of the two scalar fields product. *Computer Physics Communications*, **54**, 177-182. [7]

Martinec, Z. (1990). A refined method of recovering potential coefficients from surface gravity data. *Studia Geoph. et Geod.*, **34**, 313-326.   [1]

Martinec, Z. (1991). On the accuracy of the method of condensation of the Earth's topograhy. *Man. Geod.*, **16**, 288-294.   [5]

Martinec, Z. (1993). Effect of lateral density variations of topographical masses in view of improving geoid model accuracy over Canada. Final report under DSS contract No. 23244-2-4356/01-SS, Geodetic Survey of Canada, Ottawa, June, 1993.   [0,7]

Martinec, Z. (1994a). The density contrast at the Mohorovičič discontinuity. *Geophys. J. Int.*, **117**, 539-544.   [3]

Martinec, Z. (1994b). The minimum depth of compensation of topographic masses. *Geophys. J. Int.*, **117**, 545-554.   [5]

Martinec, Z. (1995). Numerical stability of the least-squares solution of the discrete altimetry-gravimetry boundary-value problem for determination of the global gravity model. *Geophys. J. Int.*, **123**, 715-726.   [12]

Martinec, Z. (1996). Stability investigations of a discrete downward continuation problem for geoid determination in the Canadian Rocky Mountains. *J. Geod.*, **70**, 805-828.   [1,8]

Martinec, Z. (1998). Construction of Green's function to the Stokes boundary-value problem with ellipsoidal corrections in boundary condition. *J. Geod.* (in press).   [11]

Martinec, Z., C. Matyska, E.W. Grafarend and P. Vaníček (1993). On Helmert's 2nd condensation technique. *Man. Geod.*, **18**, 417-421.   [1,4,7]

Martinec, Z. and P. Vaníček (1994a). The indirect effect of Stokes-Helmert's technique for a spherical approximation of the geoid. *Man. Geod.*, **19**, 213-219.   [1,3,5]

Martinec, Z. and P. Vaníček (1994b). Direct topographical effect of Helmert's condensation for a spherical geoid. *Man. Geod.*, **19**, 257-268.   [1,5]

Martinec, Z., P. Vaníček, A. Mainville and M. Véronneau (1995). The effect of lake water on geoidal height. *Man. Geod.*, **20**, 193-203.   [6]

Martinec, Z., P. Vaníček, A. Mainville and M. Véronneau (1996). Evaluation of topographical effects in precise geoid computation from densely sampled heights. *J. Geod.*, **70**, 746-754.   [5]

Martinec, Z. and P. Vaníček (1996). Formulation of the boundary-value problem for geoid determination with a higher-order reference field. *Geophys. J. Int.*, **126**, 219-228.   [7]

Martinec, Z. and C. Matyska (1997). On the solvability of the Stokes pseudo-boundary-value problem for geoid determination. *J. Geod.*, **71**, 103-112.   [1]

Martinec, Z. and E.W. Grafarend (1997a). Solution to the Stokes boundary-value problem on an ellipsoid of revolution. *Studia Geoph. et Geod.*, **41**, 103-129.   [9]

Martinec, Z. and E.W. Grafarend (1997b). Construction of Green's function to the external Dirichlet boundary-value problem for the Laplace equation on an ellipsoid of revolution. *J. Geod.*, **71**, 562-570. [10]

Matyska, C. (1994). Topographic masses and mass heterogeneities in the upper mantle. *Gravimetry and Space Techniques Applied to Geodynamics and Ocean Dynamics*, Geophysical Monograph **82**, IUGG vol. 17, 125-132. [1]

Milbert, D.G. (1991). Geoid90: A high-resolution geoid for the United States. *EOS Trans AGU*, **72**(49), 545. [0]

Molodenskij, M.S., V.F. Eremeev and M.I. Yurkina (1960). *Methods for Study of the External Gravitational Field and Figure of the Earth*. Translated from Russian by the Israel Program for Scientific Translations for the Office of Technical Services, U.S.Department of Commerce, Washington, D.C., U.S.A., 1962. [1]

Moon, P. and D.E. Spencer (1961). *Field Theory for Engineers*. D. van Nostrand Comp., Princeton. [9]

Moritz, H. (1966). Linear solutions of the geodetic boundary-value problem. Dept. of Geod. Sci. Rep., **79**, Ohio State Univ., Columbus. [5]

Moritz, H. (1968). On the use of the terrain correction in solving Molodensky's problem. Dept. of Geod. Sci. Rep., **108**, Ohio State Univ., Columbus. [5]

Moritz, H. (1980a). *Advanced Physical Geodesy*. H. Wichmann Verlag, Karlsruhe. [8,11]

Moritz, H. (1980b). Geodetic Reference System 1980. *Bull. Géod.*, **54**, 395-405. [9]

Moritz, H. (1990). *The Figure of the Earth: Theoretical Geodesy and the Earth's Interior*. H.Wichmann Verlag, Karlsruhe. [1,8]

Paul, M. (1973). A method of evaluating the truncation error coefficients for geoidal heights. *Bull. Géod.*, **110**, 413-425. [8]

Pellinen, L.P. (1962). Accounting for topography in the calculation of quasi-geoidal heights and plumb-line deflections from gravity anomalies. *Bull. Géod.*, **63**, 57-65. [1,4,8]

Pick, M., J. Pícha and V. Vyskočil (1973). *Theory of the Earth's Gravity Field*. Elsevier, Amsterdam. [9,10,11]

Press, W.H., B.P. Flannery, S.A. Teukolsky and W.T. Vetterling (1992). *Numerical Recipes. The Art of Scientific Computing*. Cambridge Univ.Press, Cambridge. [1,8,9,10,11]

Ralston, A. (1965). *A First Course in Numerical Analysis*. McGraw-Hill, Inc., New York. [8]

Rapp, R.H. (1986). Gravity anomalies and sea surface heights derived from a combined Geos 3/Seasat Altimetric data set. *J. Geophys. Res.*, **91**, B5, 4867-4876. [12]

Rapp, R.H. (1993). Global geoid determination. In *Geoid and Its Geophysical Interpretations*, eds. P. Vaníček and N.T. Christou, CRC Press, Boca Raton, 57-76.   [12]

Rapp, R.H., Y.M. Wang and N.K. Pavlis (1991). The Ohio State 1991 geopotential and sea surface topography harmonic coefficients models. Dept. of Geod. Sci. Rep., **410**, State Univ., Columbus.   [12]

Reigber, Ch. (1989). Gravity field recovery from satellite tracking data. In *Theory of Satellite Geodesy and Gravity Field Determination*, eds. F. Sansò and R. Rummel, Springer-Verlag, Berlin, 197-234.   [12]

Rektorys, K. (1968). *Survey of Applicable Mathematics*. SNTL, Prague (in Czech).   [8]

Rektorys, K. (1977). *Variational Methods*. Translation from Czech (1974). Reidel Co., Dordrecht-Boston.   [12]

Rummel, R. (1977). The determination of gravity anomalies from geoid heights using the inverse Stokes formula. Dept.of Geod. Sci. Rep., **269**, Ohio State Univ., Columbus.   [12]

Rummel, R. (1979). Determination of short-wavelength components of the gravity field from satellite-to-satellite tracking or satellite gradiometry - An attempt to an identification of problem areas. *Man. Geod.*, **4**, 107-148. [8]

Rummel, R., R.H. Rapp, H. Sünkel and C.C. Tscherning (1988). Comparisons of global topographic/isostatic models to the Earth's observed gravity field. Rep. **388**, Dept. of Geodetic Science and Surveying, The Ohio State University, Columbus.   [1]

Sacerdote F. and F. Sansò (1983). A contribution to the analysis of the altimetry-gravimetry problem. In *Proc. Int. Symp. Figure of the Earth, the Moon and other Planets*, ed. P. Holota, Monograph Series of VÚGTK, Prague, 122-139.   [12]

Sacerdote F. and F. Sansò (1986). The scalar boundary value problem of physical geodesy. *Man. Geod.*, **11**, 15-28.   [12]

Sacerdote F. and F. Sansò (1987). Further remarks on the altimetry-gravimetry problems. *Bull. Geod.*, **61**, 65-82.   [12]

Sailor, R.V. (1993). Signal processing techniques. In *Geoid and Its Geophysical Interpretations*, eds. P. Vaníček and N.T. Christou, CRC Press, Boca Raton, 147-185.   [12]

Sansò, F. (1981). Recent advances in the theory of the geodetic boundary value problem. *Rev. of Geophys. Space Phys.*, **19**, 437-449.   [1,12]

Sansò, F. (1993). Theory of geodetic b.v.p.s applied to the analysis of altimetry data. In *Satellite altimetry in geodesy and oceanography*, eds. R. Rummel and F. Sansò, Springer-Verlag, Berlin, 318-371.   [12]

Sansò, F. (1995). The long road from measurements to boundary value problems in physical geodesy. *Man. Geod.*, **20**, 326-344.   [0,1]

Seitz, K., B. Schramm and B. Heck (1994). Non-linear effects in the scalar free geodetic boundary value problem based on reference fields of various degrees. *Man. Geod.*, **19**, 327-338. [1]

Shaofeng, B. and D. Xurong (1991). On the singular integration in Physical Geodesy. *Man. Geod.*, **16**, 283-287. [8]

Sideris, M.G. (1994). Regional geoid determination. In *Geoid and Its Geophysical Interpretations*, eds. P. Vaníček and N.T. Christou, CRC Press, Boca Raton, 77-94. [0,6]

Sideris, M. G., I.N. Tziavos and K.P. Schwartz (1989). Computer software for terrain reductions using Molodensky's operator. Contract report, Geodetic Survey of Canada, May 1989. [0]

Sideris, M.G. and R. Forsberg (1990). Review of geoid prediction methods in mountainous regions. In *Determination of the Geoid, Present and Future*, Symposium No.106, eds. R.H.Rapp and F.Sansò, Springer-Verlag, New York, 1991, 51-62. [0,4,5,8]

Sideris, M. and B.B. She (1995). A new, high-resolution geoid for Canada and part of the U.S. by the 1D-FFT method. *Bull. Géod.*, **69**, 92-108. [0]

Sjöberg, L. (1975). On the discrete boundary value problem of physical geodesy with harmonic reductions to an internal sphere. The Royal Institute of Technology, Stockholm. [0,8]

Sjöberg, L. (1994). Techniques for geoid determination, In *Geoid and Its Geophysical Interpretations*, eds. P. Vaníček and N.T. Christou, CRC Press, Boca Raton, 33-56. [7]

Sona, G. (1995). Numerical problems in the computation of ellipsoidal harmonics. *J. Geod.*, **70**, 117-126. [10]

Stewart, M.P. and R.G. Hipkin (1990). A high resolution, high precision geoid for the British Isles. In *Sea Surface Topography and the Geoid*, eds. H. Sünkel and T.F. Baker, IAU Symposium 104, Springer Verlag, New York, 39-46 [0]

Stokes, G.G. (1849). On the variation of gravity on the surface of the Earth. *Trans. Cambridge Phil. Soc.*, **8**, 672-695. [9,11]

Svensson, S.L. (1983). Solution of the altimetry-gravimetry problem. *Bull. Geod.*, **57**, 332-353. [12]

Svensson, S.L. (1988). Some remarks on the altimetry-gravimetry problem. *Man. geod.*, **13**, 63-74. [12]

Tarantola, A. (1987). *Inverse Problem Theory*. Elsevier, Amsterdam. [12]

Telford, W.M., L.P. Geldart, R.E. Sheriff and D.A. Keys (1978). *Applied Geophysics*. Cambridge Univ. Press, Cambridge. [6]

Thong, N.C. (1993). Untersuchen zur Lösung der fixen gravimetrischen Randwertprobleme mittels sphäroidaler und Greenscher Funktionen, Deutsche Geodätische Kommission, Bayerische Akademie der Wissenschaften, Reihe C **399**, München 1993. [9,10]

Thong, N. C. and E.W. Grafarend (1989). A spheroidal harmonic model of the terrestrial gravitational field. *Man. Geod.*, **14**, 285-304.   [9]

Vaníček, P. and A. Kleusberg (1987). The Canadian geoid - Stokesian approach. *Man. geod.*, **12**, 86-98.   [0,1,3,5,6,8]

Vaníček, P., A. Kleusberg, R. Chang, H. Fashir, N. Christou, M. Hofman, T. Kling and T. Arsenault (1987). The Canadian Geoid. Dept. of Surv. Eng. Rep., **129**, Univ. of New Brunswick, Fredericton, N.B.   [7]

Vaníček, P. and L.E. Sjöberg (1991). Reformulation of Stokes's theory for higher than second-degree reference field and a modification of integration kernels. *J. Geophys. Res.*, **96 (B4)**, 6529-6539.   [0,7]

Vaníček, P., M. Najafi, Z. Martinec and L.Harrie (1995). Higher-degree reference field in the generalized Stokes-Helmert scheme for geoid computation. *J. Geod.*, **70**, 176-182.   [7]

Vaníček, P., W. Sun, P. Ong, Z. Martinec, P. Vajda and B. ter Horst (1996). Downward continuation of Helmert's gravity. *J. Geod.*, **71**, 21-34.   [1,7,8]

Varshalovich, D.A., A.N. Moskalev and V.K. Khersonskii (1989). *Quantum Theory of Angular Momentum*. World Scientific Publ., Singapore. [2,7,9,10,11]

Wagner, C.A. (1989). Summer school lectures on satellite altimetry. In *Theory of Satellite Geodesy and Gravity Field Determination*, eds. F. Sansò and R. Rummel, Springer-Verlag, Berlin, 285-334.   [12]

Wang, Y.M. (1988). Downward continuation of the free-air gravity anomalies to the ellipsoid using the gradient solution, Poisson's integral and terrain correction - numerical comparison and computations. Rep. **393**, Dept. of Geodetic Science and Surveying, The Ohio State University, Columbus. [8]

Wang, Y.M. (1990). The effect of topography on the determination of the geoid using analytical downward continuation. *Bull. Géod.*, **64**, 231-246.   [8]

Wang, Y.M. and R.H. Rapp (1990). Terrain effects on geoid undulation computations. *Man. Geod.*, **15**, 23-29.   [3,4,5]

Wichiencharoen, C. (1982). The indirect effects on the computation of geoid undulations. Rep. **336**, Dept. of Geodetic Science and Surveying, The Ohio State University, Columbus.   [1,3,5]

Wieser, M. (1987). The global digital terrain model TUG87. Internal Report on Set-up, Origin and Characteristics, Inst. of Math. Geodesy, Technical Univ. of Graz, Austria.   [1,2,7,12]

Wilkinson, J.H. (1965). *The Algebraic Eigenvalue problem*. Clarendon Press, Oxford.   [8,12]

Zlotnicki, V. (1993). The geoid from satellite altimetry. In *Geoid and Its Geophysical Interpretations*, eds. P. Vaníček and N.T. Christou, CRC Press, Boca Raton, 95-110.   [12]

Index

Springer
and the
environment

At Springer we firmly believe that an international science publisher has a special obligation to the environment, and our corporate policies consistently reflect this conviction.
We also expect our business partners – paper mills, printers, packaging manufacturers, etc. – to commit themselves to using materials and production processes that do not harm the environment. The paper in this book is made from low- or no-chlorine pulp and is acid free, in conformance with international standards for paper permanency.

Springer

Lecture Notes in Earth Sciences

Vol. 37: A. Armanini, G. Di Silvio (Eds.), Fluvial Hydraulics of Mountain Regions. X, 468 pages. 1991.

Vol. 38: W. Smykatz-Kloss, S. St. J. Warne, Thermal Analysis in the Geosciences. XII, 379 pages. 1991.

Vol. 39: S.-E. Hjelt, Pragmatic Inversion of Geophysical Data. IX, 262 pages. 1992.

Vol. 40: S. W. Petters, Regional Geology of Africa. XXIII, 722 pages. 1991.

Vol. 41: R. Pflug, J. W. Harbaugh (Eds.), Computer Graphics in Geology. XVII, 298 pages. 1992.

Vol. 42: A. Cendrero, G. Lüttig, F. Chr. Wolff (Eds.), Planning the Use of the Earth's Surface. IX, 556 pages. 1992.

Vol. 43: N. Clauer, S. Chaudhuri (Eds.), Isotopic Signatures and Sedimentary Records. VIII, 529 pages. 1992.

Vol. 44: D. A. Edwards, Turbidity Currents: Dynamics, Deposits and Reversals. XIII, 175 pages. 1993.

Vol. 45: A. G. Herrmann, B. Knipping, Waste Disposal and Evaporites. XII, 193 pages. 1993.

Vol. 46: G. Galli, Temporal and Spatial Patterns in Carbonate Platforms. IX, 325 pages. 1993.

Vol. 47: R. L. Littke, Deposition, Diagenesis and Weathering of Organic Matter-Rich Sediments. IX, 216 pages. 1993.

Vol. 48: B. R. Roberts, Water Management in Desert Environments. XVII, 337 pages. 1993.

Vol. 49: J. F. W. Negendank, B. Zolitschka (Eds.), Paleolimnology of European Maar Lakes. IX, 513 pages. 1993.

Vol. 50: R. Rummel, F. Sansò (Eds.), Satellite Altimetry in Geodesy and Oceanography. XII, 479 pages. 1993.

Vol. 51: W. Ricken, Sedimentation as a Three-Component System. XII, 211 pages. 1993.

Vol. 52: P. Ergenzinger, K.-H. Schmidt (Eds.), Dynamics and Geomorphology of Mountain Rivers. VIII, 326 pages. 1994.

Vol. 53: F. Scherbaum, Basic Concepts in Digital Signal Processing for Seismologists. X, 158 pages. 1994.

Vol. 54: J. J. P. Zijlstra, The Sedimentology of Chalk. IX, 194 pages. 1995.

Vol. 55: J. A. Scales, Theory of Seismic Imaging. XV, 291 pages. 1995.

Vol. 56: D. Müller, D. I. Groves, Potassic Igneous Rocks and Associated Gold-Copper Mineralization. 2nd updated and enlarged Edition. XIII, 238 pages. 1997.

Vol. 57: E. Lallier-Vergès, N.-P. Tribovillard, P. Bertrand (Eds.), Organic Matter Accumulation. VIII, 187 pages. 1995.

Vol. 58: G. Sarwar, G. M. Friedman, Post-Devonian Sediment Cover over New York State. VIII, 113 pages. 1995.

Vol. 59: A. C. Kibblewhite, C. Y. Wu, Wave Interactions As a Seismo-acoustic Source. XIX, 313 pages. 1996.

Vol. 60: A. Kleusberg, P. J. G. Teunissen (Eds.), GPS for Geodesy. VII, 407 pages. 1996.

Vol. 61: M. Breunig, Integration of Spatial Information for Geo-Information Systems. XI, 171 pages. 1996.

Vol. 62: H. V. Lyatsky, Continental-Crust Structures on the Continental Margin of Western North America. XIX, 352 pages. 1996.

Vol. 63: B. H. Jacobsen, K. Mosegaard, P. Sibani (Eds.), Inverse Methods. XVI, 341 pages, 1996.

Vol. 64: A. Armanini, M. Michiue (Eds.), Recent Developments on Debris Flows. X, 226 pages. 1997.

Vol. 65: F. Sansò, R. Rummel (Eds.), Geodetic Boundary Value Problems in View of the One Centimeter Geoid. XIX, 592 pages. 1997.

Vol. 66: H. Wilhelm, W. Zürn, H.-G. Wenzel (Eds.), Tidal Phenomena. VII, 398 pages. 1997.

Vol. 67: S. L. Webb, Silicate Melts. VIII. 74 pages. 1997.

Vol. 68: P. Stille, G. Shields, Radiogenetic Isotope Geochemistry of Sedimentary and Aquatic Systems. XI, 217 pages. 1997.

Vol. 69: S. P. Singal (Ed.), Acoustic Remote Sensing Applications. XIII, 585 pages. 1997.

Vol. 70: R. H. Charlier, C. P. De Meyer, Coastal Erosion – Response and Management. XVI, 343 pages. 1998.

Vol. 71: T. M. Will, Phase Equilibria in Metamorphic Rocks. XIV, 315 pages. 1998.

Vol. 72: J. C. Wasserman, E. V. Silva-Filho, R. Villas-Boas (Eds.), Environmental Geochemistry in the Tropics. XIV, 305 pages. 1998.

Vol. 73: Z. Martinec, Boundary-Value Problems for Gravimetric Determination of a Precise Geoid. XII, 223 pages. 1998.

Vol. 74: M. Beniston, J. L. Innes (Eds.), The Impacts of Climate Variability on Forests. XIV, 329 pages. 1998.

Vol. 75: H. Westphal, Carbonate Platform Slopes – A Record of Changing Conditions. XI, 197 pages. 1998.

Vol. 76: J. Trappe, Phanerozoic Phosphorite Depositional Systems. XII, 316 pages. 1998.